Praise for
The Universe Is Pranking Us

'Exhilarating, playful, broad, and deep. A spectacular survey of the frontiers of contemporary science, enriched by the dialogue between Western and Indian cultural perspectives.'

—**Dr Gregory Chaitin**

Mathematician, computer scientist, discoverer of the Omega number, and foundational contributor to Solomonoff-Kolmogorov-Chaitin Complexity

'In *The Universe Is Pranking Us*, Mrittunjoy argues that uncertainty is not just an epistemic gap but a fundamental feature of reality. Drawing from quantum mechanics, Gödel's incompleteness theorems, chaos theory, evolutionary biology, and intellectual history, the book traces how the Western quest for absolute certainty—from Plato and Descartes through Newton and Enlightenment rationalism—reaches its breaking point in modern science. The Einstein-Bohr debates and quantum theory's inherent probabilism highlight this shift, which the author juxtaposes with alternative epistemologies. The book rejects reductionism and linear progress narratives, proposing instead a relational, context-aware approach to knowledge. Far from a flaw, uncertainty becomes a generative force that calls for intellectual humility and plural perspectives.'

—**Dr Subhash Kak**

Padma Shri awardee, computer scientist, and member of the Indian Prime Minister's Science, Technology, and Innovation Advisory Council (PM-STIAC)

'Science progresses not by clinging to comforting certainties but by interrogating them. In *The Universe Is Pranking Us*, Dr Mrittunjoy Guha Majumdar revisits one of the most profound lessons of modern physics: that reality is under no obligation to conform to human intuition. This book dismantles intellectual complacency while reaffirming the primacy of rigorous inquiry. Doubt, when disciplined by reason and evidence, is not relativism; it is the very method of science.

This book is a provocative and timely reminder that humility before nature is the beginning of wisdom.'

—Dr Anand Ranganathan
Molecular biologist, bestselling author,
professor (Jawaharlal Nehru University),
and political commentator

***

By the Same Author

From Shiva to Schrödinger:
Unravelling Cosmic Secrets with
Trika Shaivism & Quantum Insights

a journey through the philosophy and physics of a reality that thrives on uncertainty

the universe is pranking us

Dr Mrittunjoy Guha Majumdar

HAY HOUSE INDIA
New Delhi • London • Sydney
Carlsbad, California • New York City

Hay House Publishers (India) Pvt Ltd
Muskaan Complex, Plot No. 3, B-2, Vasant Kunj, New Delhi – 110070, India

Hay House LLC, P.O. Box 5100, Carlsbad, CA 92018-5100, USA
Hay House UK Ltd, 1st Floor, Crawford Corner, 91-93 Baker Street, London W1U 6QQ, UK
Hay House Australia Publishing Pty Ltd, 18/36 Ralph St., Alexandria NSW 2015, Australia

Email: contact@hayhouse.co.in
Website: www.hayhouse.co.in

First published by Hay House India, 2026
10 9 8 7 6 5 4 3 2 1

ISBN 978-93-6611-933-5 (Paperback)
ISBN 978-93-6611-011-0 (Ebook)

Printed at Thomson Press India Private Limited

FSC www.fsc.org
MIX
Paper | Supporting responsible forestry
FSC® C010615

Hay House is committed to a sustainable future for our business, our readers and our planet. This book is made from Forest Stewardship Council® certified paper.

For my firstborn, Anandalekha—my ray of bliss
in a universe of beautiful uncertainty.

contents

foreword

Religious doctrine is familiar to us, but the existence of a scientific equivalent—and the fact that denying current doctrine may have serious consequences for the denier—is less well known. Jacques Benveniste, following his claim that highly diluted solutions of biological substances can retain biological properties despite containing no molecules of the original substance, discovered this to his cost, just as did the discoverer of continental drift, something climate scientists at the time considered impossible.

'Memory of water' is of particular interest to me because I was present at the conference where Benveniste first announced his discovery, before the vicious attack on him by the then editor of *Nature* (who later received a knighthood for his work). I often make the following point: the claim that water can have a memory can be readily disproved by any one of a number of easily understood but invalid arguments. An example is the argument that can be summarised as 'no molecules, no effect,' the problem being that it ignores the possibility—supported by computer simulations—that water itself may form complex molecular structures which could have their own biological activity.

An equally invalid argument, once put to me by a well-known physicist, is the claim that water changes its microscopic structure on such a short time scale that memory over any period would be impossible. Had this argument been valid, it would equally have demonstrated the impossibility of ferromagnetism. Scientists often say to one another about supposed heresies, 'Clearly this is wrong,' and this is how oversimplified perspectives persist.

The above cases involve false beliefs, and that is the focus of this book, which asks: Why are we so confident about what is inherently uncertain? Investigators have tended to presume that they are discovering the truth, but that is not actually how things are. Whatever seems true to us now may no longer hold in the future, as time passes and more is learnt about how the world works—a state of affairs explored in detail in the pages that follow.

—Professor Brian Josephson
Nobel Laureate in Physics, 1973

preface

What if everything you have been told about finding truth is wrong? Right now, someone is promising you the answer—the answer to life, the universe, and everything within. Forty-two, you say? Not quite. Some algorithm is working overtime to convince you that the world is all about sure-thingness. But what if this is just another form of absoluteitis—an affliction that makes one smug about smoke and confident about what is inherently uncertain? What if the universe gives a wry smile every time we display our adorable human need to wrap reality in neat little packages? Because while we have been busy building intellectual fortresses of absolute truth, the cosmos has been playing by entirely different rules. Rules that change mid-game. Rules that contradict themselves. Rules that are not really rules at all. And it is in this tension between what we claim to know and what reality actually does that the real story begins.

The truth is that our entire quest for ultimate truth has been magnificently, beautifully, catastrophically misguided. You have felt it, haven't you? That moment when the ground shifts beneath your feet. When the formula that worked yesterday fails today. When the expert's prediction crumbles like stale bread. When you realise that behind all our confident proclamations lies something far more mysterious and alive than any theory can capture.

The Universe Is Pranking Us is your invitation to stop running from that feeling and to understand that nature is inherently full of uncertainties and grey zones. And nowhere is this more evident than in the age we live in. Your phone knows where you will be in an hour. Your feed knows what you will want to buy. Your GPS knows the fastest route home. But strip away the digital smoke and mirrors, and you will find the same truth that has always haunted human consciousness: the universe does not do absolute. It does paradox. It does mystery. It does the kind of cosmic joke that makes quantum physicists weep and philosophers reach for stronger coffee.

Here is the secret they don't teach in school: the greatest minds in history—Einstein, Heisenberg, and Gödel—didn't triumph because they found all the answers. They triumphed because they had the courage to ask better questions. Questions that exploded the cosy certainties of their time and revealed a reality far stranger and more wonderful than anyone had imagined. And that spirit of questioning is exactly what this book hopes to revive.

Put simply, *The Universe Is Pranking Us* isn't here to give you answers. The world is drowning in answers. This book is here to give you something far more precious: the art of not knowing with grace. We are going to discover that uncertainty isn't the enemy of wisdom; it is wisdom's most intimate companion. That chaos isn't the opposite of order; it is order's dance partner. And most importantly, this book is here to show you that the question mark isn't a sign of weakness; it's the most powerful punctuation mark in the universe.

Before we begin, a fair warning: this journey may unsettle you, challenge your assumptions, and even dismantle a few 'truths'. It will make you doubt everything you think you know about knowing. It will ask you to hold contradictions without collapsing, to embrace questions without demanding immediate answers, to find your footing on ground that keeps shifting beneath your feet. But here is what you will discover on the other side: freedom. The freedom that comes from no longer needing to be right all the time. The freedom to be curious instead of certain. The freedom to dance with mystery instead of wrestling it to the ground. You will find that when you stop trying to control the universe, something far more interesting happens: you start to participate in it. You become part of the cosmic conversation instead of its self-appointed judge and jury.

All I will say is to close your certainty apps. Unsubscribe from the truth-sellers. Put down the self-help books that promise to solve your life in ten easy steps. Instead, bring your sense of wonder. Bring your willingness to be wrong. Bring your capacity to laugh at the cosmic joke of human existence and to find in that laughter something approaching wisdom. The universe is vast, mysterious, and utterly indifferent to our need for neat explanations. It is also the most magnificent playground ever conceived, full of surprises, joy, and paradoxes.

Welcome to the world of the absurd. Welcome to the wild, beautiful, impossible adventure of not knowing.

—Dr Mrittunjoy Guha Majumdar

'Nature's imagination is so much greater than man's.'

— **Richard Feynman**

chapter 1

the maybe universe

Absolutely! A word more worshipped than witnessed. We live in an age where everyone wants the mantra of certainty. Five-step formulas to success. Guaranteed morning routines. Clickbait headlines promising 'The One Truth You Need to Know.' From scientific consensus to life hacks to manifestation techniques, the hunger for absolute answers is everywhere—fast, packaged, and algorithm-approved.

The problem isn't just the abundance of answers, but what their absence now seems to imply. In a world addicted to quick fixes, doubt feels like failure. Every TED Talk, tweet, and self-help book whispers: Be sure. Be right. Be done. There is a bestseller waiting for anyone who can slap 'guaranteed' on a cover—guaranteed love, guaranteed money, guaranteed enlightenment by Friday. Only if I could have put 'guaranteed uncertainty' on this book!

The truth is that today's algorithm gods favour those who sound sure as nuance doesn't trend and humility doesn't get likes.

That which is loud, sensationalist, out there, sells. Sometimes, it almost feels like we are one downloadable treatise away from solving the ultimate mysteries of the universe. Or so Instagram thinks. And once certainty becomes a performance, it doesn't take long for the stage to fill up. This is the reason why everyone is a guru now. If you doubt it, just check their bio: 'Truth Coach. Life Strategist. Cosmic CEO.'

Beneath the glitter and dopamine rush of 'knowing for sure,' something quietly rots. The very ground we stand on—science, philosophy, even reason—has been whispering for a century now that certainty is a beautiful lie. The Age of Enlightenment promised predictable precision, only for quantum physics to later shatter its foundations. Philosophy, too, aimed for rational mastery. However, now it stutters between infinite perspectives, like a radio stuck between stations. But have we ever asked ourselves: what if within reality, there isn't an answer to be found? What if the highest wisdom is not knowing? What if the purest knowledge is learning how to lose your grip and still sing?

But before we start celebrating uncertainty, let's take a step back and try to understand this obsession with certainty. In fact, this is not just a modern problem; it is a centuries-old quest. The Western intellectual tradition has been hell-bent on finding solid ground, as if the universe could be captured in neat, tidy packages with each label reading: 'Objective truth. Verified fact. Irrefutable law.' But what if all this intellectual architecture, built brick by brick, is more like a sandcastle at high tide, destined to crumble under the weight of a deeper truth? Maybe the illusion of absolutes has been a comfort blanket we have wrapped ourselves in, instead of a map to the actual terrain. And it turns out that the more we try to pin reality down, the more it slips through our fingers. Think of Plato's eternal forms—perfect, unchanging, forever out of reach. Descartes' methodical doubt in search of a foundation so solid it could withstand even the most vicious scepticism led him to the cogito—the mind's unshakeable certainty of itself. Then Newton came along, laying down the clockwork universe, where everything had its place and its cause, where certainty ruled with an iron hand. And yet, this too unravelled when Einstein waltzed in, telling us that time wasn't a fixed backdrop, but something that bent, warped, and danced with mass and energy. Then came quantum physics and suddenly, everything we thought we knew about the universe started to fall apart. It is important to note here that quantum mechanics didn't throw away the rules; it just changed them. Instead of everything being predictable, we got entanglement and superposition, where particles can be in more than one state at the same time. Heisenberg's uncertainty principle tells us that we can never know both a particle's position and speed exactly at the same time as uncertainty is built into how reality works. Then, Gödel stepped in and showed that no system, whether in math or logic, can prove all truths on its own. All these breakthroughs revealed that some things will always be beyond our ability to prove.

Western philosophers like Nietzsche and Derrida, who challenged the idea of a fixed, universal truth, have become iconic figures in modern thought. We have all heard Nietzsche's provocative claim that 'God is dead,' which has since become a cultural product, appearing on everything from T-shirts to tattoos. However, Nietzsche also famously proclaimed, 'There are no facts, only interpretations,' a notion that captures the messy reality of a world shaped by competing narratives and perspectives. Nietzsche and Derrida both basically called out the human tendency to chase certainty, saying, 'Hey, we are missing the bigger picture!' They reminded us that our knowledge is always a bit like trying to assemble Ikea furniture without all the parts; no matter how much we try to force it, there are always cracks. Whether it is science or philosophy, the big ideas we have built are just a snapshot from where we are standing, and they are bound to have a few wobbly bits when we look closer. However, it is worth noting that while Nietzsche and Derrida challenged the idea of absolute truth, they still operated within the traditional subject-object divide, which they sought to dismantle but never fully escaped. This lingering separation between observer and observed would soon face a far more radical challenge from an unexpected direction.

It was then that John von Neumann entered the picture, not only laying the foundations of modern computing but also grappling with the quantum conundrum of the role of the observer. Could we ever separate the observer from the observed, or are they forever tangled like your phone's charging cable at the bottom of your bag? Around the same time, Alan Turing wasn't just cracking codes during World War II, but also defining the limits of computation, setting up an ultimate 'no entry' sign for certain classes of problems. Meanwhile, philosophers like Ludwig Wittgenstein were wrestling with language and its limits, suggesting that the boundaries of our language were the boundaries of our world. In other words, if you can't put something into words, understanding it becomes a serious challenge. And while all this intellectual heavy-lifting was going on, the French mathematician Henri Poincaré was declaring, 'I never believe anything I cannot prove.' But even he wasn't immune to the cracks in the absolute. He argued that mathematics rests on conventions, a blend of structure and creativity, shaped as much by what we choose to define as by what is inherently true.

The irony is that while these thinkers were each immersed in their own intellectual worlds, they were all, unknowingly, moving towards the same conclusion: the search for certainty is not as certain as we once thought. The universe, it turns out, is not a perfectly oiled machine but something closer to a quantum funhouse, full of unexpected twists, strange reflections, and mirrors that make you wonder whether you are looking at the world or at yourself.

Can 'Eureka!' Be by Design?

The Western hunt for absolute truth began in ancient Greece. Plato was the first great superhero of Western metaphysical thought, imagining a realm of perfect forms, i.e., unchanging ideals like justice. Think of it as the cloud storage of all ideal concepts that are eternal, pristine, and untouched by the messiness of real life. This sentiment finds its clearest expression in his famous allegory of the cave, where we are prisoners watching flickering shadows on a wall and mistaking them for reality. The philosopher is the one who breaks free, steps into the light, and finally sees truth for what it truly is.

Then comes Aristotle, Plato's more pragmatic sequel. If Plato was the dreamer, Aristotle was the builder—more Tony Stark than Doctor Strange! He had little interest in a hidden realm of perfect forms. Instead, he looked at the world and said, 'Let's start with what we can observe, measure, and dissect.' He classified animals, analysed causality, and laid the foundations of science itself. Yet even he believed in universal principles that could be discovered through reason, like timeless coding in the source code of the cosmos. Together, they fuelled the West's obsession with certainty, a bit like building a Google Maps for truth that is neat, absolute, and always steering you towards the 'one right answer.' But what if reality is less GPS and more Waze, shaped by context, constantly updating, and prone to throwing in the occasional unexpected detour?

The story takes a fascinating turn when we move from the lofty dialogues of Plato to the clean lines of mathematics and geometry. Ancient Greece wasn't just philosophising about the soul; it was also sketching blueprints for the cosmos. Pythagoras, almost a mystic mathematician, believed numbers were not merely tools but sacred keys to the universe. For him, harmony wasn't metaphorical; it was literal. He believed that music, planets, and even the human soul could be expressed through precise numerical ratios.

Then came Euclid, laying out his geometric postulates with the clarity of a poet writing in axioms. In *Elements*, the world became a canvas of triangles, circles, and proofs—a place where certainty could be drawn with a compass and straightedge. Truth here was eternal, like an untouched mandala of logic. This passion for permanence proved so powerful that it shaped Western science for millennia. Newton's laws echoed the same longing: timeless truths etched into the structure of reality, waiting to be uncovered, much like the Pythagorean harmony of the spheres.

Yet there was always a quieter voice in the background that questioned the permanence of anything. Heraclitus was that voice. He famously said:

> 'All things are in flux; the flux is subject to a unifying measure or rational principle. This principle (logos, the hidden harmony behind all change) bound opposites together in a unified tension, which is like that of a lyre, where a stable, harmonious sound emerges from the tension of the opposing forces that arise from the bow bound together by the string.'

The most crucial takeaway is that, in a world obsessed with absolutes, he dared to insist that everything flows. Reality, for him, was not a fixed map but a living current. This insight is captured in his famous remark, 'You cannot step into the same river twice.' It was a poetic rebellion against the idealism of Plato's forms and the assumed solidity of Euclidean space.

What makes this insight even more intriguing is that it does not stand alone in the history of ideas. In fact, the story begins to sound unmistakably Indian. In the Upanishadic vision, the world is not false, yet neither is it ultimate—a shifting appearance grounded in a deeper reality. In Buddhism, everything is *anicca*—impermanent and conditioned. It is almost as if Heraclitus had wandered into a *sabha* (assembly) of Indian *rishis* (sages), nodding along as they dismantled certainty with compassion, nuance, and metaphysical patience.

We then encounter the sophists, the wandering intellectuals of ancient Greece who challenged the philosophical order long before Nietzsche or Derrida gave such challenges their modern flair. To the strict disciples of Plato and Aristotle, they were troublemakers; to others, they were early pioneers of perspective. Chief among them was Protagoras, who boldly declared, 'Man is the measure of all things.' In that single line, he cracked the marble pedestal on which universal truth was made to stand.

For Protagoras and his fellow sophists, truth was not a divine form floating in some metaphysical heaven; it was local, lived, and deeply human. It shifted with perception and moved with the rhythms of culture. Moreover, the sophists weren't cloistered thinkers. They were public intellectuals who taught rhetoric and persuasion to aspiring statesmen in the bustling marketplaces of Athens. One might think of them as the ancient world's debate coaches, TED-style mentors, and political strategists combined, sharpening arguments the way craftsmen prepared chariots for battle.

To their critics, including the famously confrontational Diogenes, they were little more than merchants of cleverness, selling verbal tricks in place of genuine wisdom. Yet even as they were mocked as con artists of the mind, their insistence on the contingency of knowledge laid important groundwork for later debates on relativism, ethics, and power. In their world, language did not merely describe reality; it actively shaped it. Like *kathakars* (storytellers) in an epic, they understood that who tells the story, and how, often determines what is accepted as true.

In a broader cultural sense, their thinking mirrors Indian philosophies such as Jainism's *anekāntavāda*, which holds that reality is too complex to be captured from a single point of view. Hence, where Plato searched for eternal ideals, the sophists turned towards the agora, with all its noise, argument, and lived complexity. In doing so, they forced civilisation to confront a lasting question: What if truth is not a single flame, but many flickering lamps, each casting its own shadow?

It is also worth noting that the ancient Greeks were not merely pondering abstract puzzles; they were setting the stage for a philosophical tug-of-war that still shapes our understanding of truth today. On one side stood the absolutists, figures like Plato and Euclid, who dreamt of eternal forms and geometric perfection. They sought to pin down truth the way one maps constellations in the night sky: fixed, orderly, and timeless. On the other side were the relativists, the sophists with their persuasive rhetoric and Heraclitus with his ever-flowing river, who saw reality as a shifting mosaic, too fluid to be captured by rigid definitions.

This tension between the eternal and the ephemeral, between the Olympian urge for absolutes and the restless sea of human perception, did more than define Greek thought; it became part of the DNA of Western philosophy. And it was never a purely academic quarrel. The debate echoed through history, reappearing in new guises. When Newton observed falling apples and orbiting moons, he envisioned a universe governed by certainty, not unlike the concept of *ṛta* in Vedic cosmology: ordered, predictable, and absolute. His laws became the new scientific sutras.

Yet just as Vedanta's idea of Māyā questions the permanence of the world, Einstein's theory of relativity unsettled Newton's clockwork universe. Time could stretch, space could curve, and reality revealed itself to be more subtle than previously imagined. Quantum mechanics then pushed the rupture even further. Determinism gave way to probability, particles existed in superposition, and certainty collapsed the moment observation entered the frame.

In this scientific upheaval, the ancient Greek debate between fixed truths and contextual realities found new life. The sophists, once dismissed as intellectual provocateurs, began to sound uncannily like modern physicists and postmodern thinkers. And Heraclitus might as well have been speaking of quantum fields rather than rivers. Even in our digital age, where AI systems evolve with new data and truth is filtered through algorithms and cultural lenses, the ancient dialectic between universalism and perspectivism continues to pulse beneath our screens. What once unfolded in the agora now reappears in laboratories and code, reminding us that the question of how truth holds—or shifts—has never truly been settled.

The Cartesian Shake-up

In the aftermath of Europe's religious wars and scientific turmoil, Descartes arrived with a philosophical game-changer: *Cogito, ergo sum* (I think, therefore I am). Surrounded by uncertainty, he went looking for a foundation for knowledge so solid that it could not be doubted. In effect, he was pressing the ultimate intellectual reset button. This was not merely about proving existence; it was about rebuilding knowledge from the ground up, beginning with a single indubitable truth: I think, therefore I exist.

Descartes' bold move was to privilege 'clear and distinct ideas' over the unreliable testimony of the senses. Sight, sound, and touch, he argued, were as deceptive as a magician's tricks. From this move emerged the famous Cartesian split: the divide between mind and body. Thought and consciousness belonged to one realm; the physical world belonged to another, governed by predictable and mechanical laws. The universe, in Descartes' vision, was a vast, well-oiled machine operating according to precise mathematical rules, like clockwork in motion. This mechanistic outlook was not merely philosophical.

Descartes also introduced the Cartesian coordinate system, a grid of X and Y axes that allowed reality to be mapped, measured, and mastered. More than a mathematical tool, it became a metaphor for reason's authority over chaos. In this sense, Descartes was not just a philosopher or mathematician; he helped lay the intellectual foundations of modern science by equipping it with methodical doubt, mathematical certainty, and the conviction that the world could be fully understood through reason. Yet, intriguingly, Descartes was no great admirer of the education system of his time. In fact, he famously remarked, 'I found myself beset by so many doubts and errors that I came to think I had gained nothing from my attempts to become educated but increasing recognition of my ignorance.'

This disillusionment with formal learning pushed Descartes to question almost everything and set him on a search for an indubitable foundation for knowledge. His philosophical project was not about passively inheriting what others had already concluded; it was about stripping away layers of uncertainty and rethinking how we arrive at truth. The challenge he set himself was stark and uncompromising: doubt everything until something beyond doubt finally reveals itself. But here lies the twist. While Descartes believed that reason and mathematics could unlock the universe's deepest mysteries, his method does not always hold up in practice. As we will see, there are many domains where reason alone proves insufficient.

In quantum mechanics, for instance, particles do not follow neat, deterministic paths. They exist in probabilistic states, directly undermining Descartes' vision of a fully mechanistic, clockwork universe. Chaos theory delivers a similar blow. It shows that even simple systems can behave in wildly unpredictable ways, where tiny differences in initial conditions produce dramatically different outcomes. Under such conditions, long-term prediction becomes impossible, no matter how refined the mathematics. Thus, we can see how Descartes' ideal of a perfectly ordered universe begins to unravel when confronted with quantum uncertainty and chaotic behaviour. But the tension doesn't end there. His famous dualism, the separation of mind and body, also fails to capture the full picture. By treating the mind as a detached observer of the body, Descartes left little room for the insights of neuroscience and psychology, which now show that thought is deeply entwined with bodily experience, emotion, and instinct.

Modern research has shown that human consciousness and decision-making cannot be explained by reason alone. Something more complex is always at work: intuition, unconscious bias, and the untidy, irrational aspects of life that resist neat Cartesian frameworks. But here's a curious twist. Descartes was not only devoted to logic and science; he was also fascinated by the pineal gland, that small and mysterious structure in the brain. He believed it to be the point where mind and body meet. The idea became his personal solution to the mind-body problem, a kind of mental portal linking thought and matter. Later thinkers, however, found the idea deeply problematic, and the pineal gland eventually lost its status as the supposed bridge between thought and matter.

While his ideas about the pineal gland never became central to modern bioscience, Descartes' influence on physics was profound. His vision of the universe as a vast mechanistic system, in which everything from planets to falling stones followed mathematical laws, set the stage for thinkers who came after him. A few decades later, Isaac Newton would adopt this mechanical outlook and apply it to the cosmos, giving us his laws of motion and gravity. In this sense, Descartes laid the intellectual groundwork not only for psychology but also for the physics that would shape our understanding of the universe.

However, not everyone accepted Descartes' dualism, the sharp division between mind and body. Baruch Spinoza rejected it outright, arguing for a holistic view in which mind and matter were inseparable expressions of a single, infinite substance. Leibniz was equally unconvinced. He proposed his own vision of reality composed of monads, tiny indivisible units coordinated in harmony, offering an alternative to Descartes' rigid separations. Yet despite these critiques, Descartes' idea of a mathematically ordered universe and a rational mind capable of understanding it became the intellectual backbone of modern science. Additionally, Descartes also stands as an intellectual forebear of computer science. His work in analytical geometry laid the foundations for the algorithms and coordinate systems that underpin our digital world today.

To conclude, Descartes did more than separate mind from body. He created the conditions in which modern science could flourish, giving us the confidence to think systematically, measure precisely, and exert control over nature. Without him, we might still be lost in medieval mysticism, navigating the world without coordinates, or worse, trying to compute the meaning of life through alchemy and astrology.

Yet while Descartes laid the groundwork for modern science through methodical doubt and faith in mathematical certainty, his vision of a fully rational, mechanistic universe begins to falter when faced with the unpredictability of quantum mechanics, chaos theory, and lived human experience.

As we move forward, it will become clear that understanding sometimes demands more than logical structures. It requires the flexibility to live with uncertainty, complexity, and even the unknown. Descartes' search for certainty gave us a powerful starting point, but it also revealed the limits of reason in our attempt to understand the world.

Detaching Observers from a Mechanistic Universe

As the 17th century unfolded, humanity's quest to understand the heavens reached a turning point, and Isaac Newton emerged as the mind capable of charting the universe with unparalleled clarity. His synthesis of terrestrial and celestial mechanics was more than a scientific breakthrough; it was the intellectual equivalent of humanity installing its first cosmic navigation system. With his laws of motion and universal gravitation, Newton presented a universe that operated with the elegance and regularity of an immense celestial clock. Put simply, the cosmos was no longer a capricious or unknowable expanse. It became a vast, decipherable puzzle in which the path of every falling apple and orbiting planet could, at least in principle, be predicted once the relevant variables were known.

Equally important is the fact that Newton's mechanical cosmos reshaped not only what we knew about the universe, but also how knowledge itself was supposed to be acquired. In this new vision, the ideal scientist was not an active participant in the drama of existence but a detached and impartial observer. Understanding nature required separating facts from values, observation from interpretation, and sensation from judgement. The emerging scientific ethos demanded what might be called epistemic asceticism—a disciplined effort to strip away personal, cultural, and emotional influence in the pursuit of objective truth. The ideal became the so-called 'view from nowhere,' knowledge gathered as if from no particular place and untouched by human limitation. Science seemed to ask the knower not merely to 'know thyself,' but to transcend the self altogether.

This model of objectivity, although remarkably successful in many domains, was deeply shaped by the social, political, and religious currents of its time. In Protestant Europe, where the Reformation had emphasised individual scrutiny and direct access to truth, this approach fit neatly into broader cultural patterns. Nature, like scriptures, was to be read plainly and precisely, without intermediaries or imaginative embellishment. The mechanical philosophy also aligned with the early modern fascination with automata. Such intricate devices mimicked life itself. At the court of Louis XIV, mechanical gardens featured animated statues and elaborate fountains that seemed to embody the spirit of a mechanised cosmos.

René Descartes, Newton's philosophical predecessor, had even suggested that animals were nothing more than complex machines, automata without consciousness and governed entirely by physical laws. The claim was so extreme that Descartes is said to have performed anatomical experiments on live animals under the belief that they were incapable of true suffering. Yet Newton himself was far from the purely rational figure he is often imagined to be. Less widely known is the fact that he devoted more time to alchemy and biblical prophecy than to physics or optics. His fascination with scripture is evident in *Observations upon the Prophecies of Daniel and the Apocalypse of St. John*, in which he attempted to decode apocalyptic visions with mathematical precision. He once wrote, 'The first time I set myself to the study of the holy scriptures was when I was about seventeen years old,' revealing a lifelong obsession with uncovering hidden meanings in the Bible. Therefore, Newton's belief in universal laws was driven by something close to religious devotion. He sought not only to describe nature but to uncover the divine blueprint behind creation. In this sense, the clockwork universe was never entirely secular. It was animated by the conviction that God was the master watchmaker, who had wound the universe and set it in motion according to laws waiting to be discovered.

The impact of this mechanical and detached view of knowledge was immense. It helped propel the Industrial Revolution, giving humanity unprecedented power to engineer its environment. Railways, steamships, and later factories emerged, their precise rhythms mirroring the imagined regularity of nature itself. Scientific knowledge became a form of power, and objectivity rose to the status of a moral ideal. However, this triumph came at a significant cost. By elevating detached, mechanistic reasoning as the only legitimate path to truth, other ways of relating to the world were marginalised. Emotional, intuitive, aesthetic, and spiritual modes of knowing were increasingly dismissed as irrational. Indigenous knowledge systems, oral traditions, and holistic sciences that emphasised interdependence and relationality were pushed aside as primitive or unscientific. Nature was no longer seen as a living partner in dialogue but as a passive resource to be analysed, exploited, and controlled.

In this way, the Newtonian legacy shaped not only modern science but a particular vision of modernity itself. It privileged mastery over participation, control over communion, and prediction over wonder. We built vast cities of knowledge on a single blueprint and, for a time, forgot that other architectures of understanding existed.

As we shall see, the challenges to this vision did not take long to arrive. Quantum mechanics, chaos theory, ecological science, and contemporary philosophy would later expose the cracks in the structure Newton helped erect. Together, they invite us to rethink what it means to know, to observe, and to participate in the unfolding of the cosmos.

Cultural Quest for Certainty from Flux

As we have covered earlier, Western civilisation has long been captivated by the idea of absolutes, and this is nowhere more evident than in its understanding of history as a linear and irreversible march towards a single destination. From the Christian expectation of Judgement Day to Hegel's vision of spirit unfolding through time, from Marx's promise of a classless society to the technology-driven utopias of Silicon Valley, time has been imagined as an arrow shot towards a fixed end where reason, progress, or salvation ultimately await.

This narrative of forward momentum has deeply shaped attitudes towards the past and the future. The past is often treated as primitive, provisional, or embarrassingly incomplete, while the future is cast as the moment of final fulfilment, the point at which humanity realises its supposed purpose. The Enlightenment gave this story institutional force by framing history as a steady ascent from superstition to science, from tyranny to freedom, and from ignorance to knowledge.

Immanuel Kant captured this spirit in his 1784 essay *What Is Enlightenment?*, where he defined enlightenment as humanity's emergence from self-imposed immaturity. Immaturity, he argued, was not a lack of intelligence but a failure of courage and the reluctance to think without guidance. His rallying cry, '*Sapere aude* (dare to know),' became the moral slogan of the age and transformed historical progress from a descriptive idea into a prescriptive demand.

To move forward was no longer just desirable; it became an ethical obligation. Those who did not fit neatly into this trajectory—indigenous peoples, traditional cultures, and even the slow, cyclical rhythms of the natural world—were dismissed as remnants of an earlier stage of development. They were seen as obstacles to be corrected or overcome in the name of progress. Colonialism cloaked its violence in the language of improvement and benevolence, claiming to usher 'backward' societies into the inevitable stream of human advancement. Even the devastation it left behind was justified as a necessary cost of development.

Remarkably, this faith in linear progress has survived the profound ruptures of the twentieth century—two world wars, genocides, the threat of nuclear annihilation, and the beginnings of ecological collapse. It continues to resurface with every technological breakthrough, every political promise to 'build a better future,' and every product marketed as smarter, faster, and more advanced. The vocabulary may change, but the underlying conviction remains intact: history is still imagined as an unwavering march towards a predetermined future.

Alongside the obsession with a fixed future emerged a parallel fixation on reductionism, the belief that to truly understand any system one must dismantle it, dissect it, and isolate its smallest parts. In this view, nature itself became a machine, ready to be taken apart like a clock, with the assumption that understanding each cog and gear would reveal the logic of the whole. This approach has undeniably borne fruit. From Newton's laws of motion to the breakthroughs of quantum mechanics, from the mapping of human anatomy to the depths of genomics, the modern world has been shaped by minds trained to simplify complexity into discrete, knowable units.

Yet for all its successes, reductionism conceals important limitations. Complex living systems do not behave like machines. Bodies, ecosystems, and societies generate emergent properties, feedback loops, and sudden shifts that resist tidy explanation. A medicine that ignores the patient's story, a psychology that reduces human experience to neurotransmitters, or an economic theory that elevates isolated rational actors while communities unravel all reveal the cost of treating wholes as nothing more than the sum of their parts.

Nature itself offers a quiet rebuttal. Pick up a pinecone and count the spirals on its scales. You may find eight winding in one direction and thirteen in the other, or perhaps thirteen and twenty-one. These numbers belong to the Fibonacci sequence, in which each term is the sum of the two preceding it. This pattern is not accidental but reflects phyllotaxis, a process through which plants organise growth efficiently. Order, in other words, emerges naturally from complexity rather than being imposed from outside. The same principle appears in a snowflake's symmetry or in the balanced branching of a tree, where structure arises from dynamic, living processes.

In plain terms, evolution produces hybrids and anomalies that defy rigid categories. Human societies are even more fluid, generating identities and affiliations that cross boundaries and change over time. Despite this, classificatory systems persisted, often with damaging consequences. Racial hierarchies were used to justify slavery and colonial domination. Economic models froze complex societies into static categories such as developed and underdeveloped. These frameworks, born of reductionist thinking, simplified reality at the cost of justice and understanding.

The reductionist drive for mastery through dissection was closely tied to another cultural impulse: the urge to classify, categorise, and impose order on the world. From Aristotle's attempts to catalogue all living beings to Linnaeus's system of binomial nomenclature, Western thought has sought to tame nature by naming it. To label something, whether a species, a mental disorder, or a civilisation, was to assert control over it.

A striking illustration of this limitation emerged in the 1960s with the appearance of koro syndrome in parts of Southeast Asia. Men experienced an intense fear that their genitals were shrinking and might retract into the body, leading to death. Western clinicians initially dismissed the phenomenon as mass hysteria. Anthropologist Arthur Kleinman, however, approached it differently. By examining the cultural context, he showed that koro was deeply entwined with local beliefs about masculinity, sexuality, and social anxiety. It could not be understood as a universal psychiatric disorder divorced from its cultural setting. Kleinman's work challenged reductionist assumptions in mental health and underscored a crucial insight: meaning, context, and culture are not optional extras but central to understanding human experience.

Thus, it becomes evident that beneath the ambition to define everything, whether through linear progress, reductionism, or the classification of existence into neat compartments, lies a universal yearning: the search for certainty in the midst of chaos. This impulse is natural and has driven remarkable achievements. It has taken us to the stars through telescopes, cured diseases through vaccines, and raised entire cities from bare ground. Yet in pursuing this certainty, we have often lost sight of worlds that resist such tidy narratives, worlds where time is not a straight arrow, identities remain fluid, and systems intertwine rather than simplify.

Many ancient traditions offer a different way of understanding existence. In Daoism, the principle of *wú wéi*, often translated as non-action, encourages alignment with the natural flow of the universe rather than the imposition of rigid control. The world, in this view, is not a puzzle to be solved but a dance to be joined. The yin-yang symbol expresses this sensibility clearly: opposing forces are not adversaries but complementary aspects of a single, dynamic whole.

A similar openness appears in Indic philosophies, where life is understood as *līlā*, a divine play. Here, the cosmos resists fixed classifications and invites participation in an unfolding mystery rather than submission to a final explanation. Knowledge is not about closure but about engagement with a living process. This relational understanding is also central to many African worldviews, particularly in the concept of *ubuntu*, often expressed as 'I am because we are.' Identity, from this perspective, does not exist in isolation but emerges through relationships. The individual cannot be separated from the community, and the whole always exceeds the sum of its parts. Unlike the Western image of the self as a bounded unit, ubuntu emphasises a dynamic, interconnected existence.

Ultimately, despite the depth of insight found across these traditions, the Western hunger for absolutes persists and often obscures the unruly richness of lived reality. We have built powerful instruments of knowledge, from telescopes and microscopes to towering skyscrapers, but in doing so, we have sometimes overlooked a fundamental truth. Therefore, we need to recognise that life does not submit easily to rigid categories. It behaves less like a machine and more like a river, constantly moving, reshaping itself, and indifferent to the maps we impose upon it. Perhaps wisdom lies not in commanding the river's course but in learning how to move with its current, to remain attentive, flexible, and alive to the flow of existence itself.

Rethinking Dynamics in a World of Complexity

Modern economics has given us a curious character: *Homo economicus,* the idealised, hyper-rational individual who always makes decisions by carefully calculating the best possible outcome. This tidy figure, shaped in the nineteenth century by economists such as William Stanley Jevons, Carl Menger, and Léon Walras, was meant to simplify economic life by making markets more predictable.

As an abstraction, Homo economicus served a purpose. It functioned much like the frictionless plane in physics: useful for building models and solving equations, but ill-suited to describing how real people actually behave. Walras himself acknowledged this limitation. His model of general equilibrium, in which all markets balance perfectly, was never intended as a literal picture of the world but as a theoretical device.

Over time, however, this abstraction hardened into something closer to a belief. Homo economicus evolved into the cultural assumption that human beings are, at their core, rational maximisers of self-interest. This idea seeped beyond economics into law, politics, and social theory, shaping how we think about voters, consumers, and even moral responsibility. Furthermore, in the 1950s, the French economist Maurice Allais exposed a crack in this logic with what later became known as the Allais paradox. He demonstrated that even well-educated, numerate individuals routinely make choices that violate the very rules of rational decision-making economists took for granted. This was not entirely new. As early as the eighteenth century, the St. Petersburg paradox, analysed by Daniel Bernoulli, revealed that people reject gambles with extremely high expected value, suggesting that human decision making does not align neatly with pure probabilistic logic.

These anomalies laid the groundwork for behavioural economics, a field devoted to understanding how people actually decide rather than how they are supposed to decide. Here, it is worth noting that emotions, habits, social pressures, and cognitive biases all play a role. And the result is a picture of human decision-making that is far messier than the elegant models suggest. In this light, Homo economicus begins to look less like a hidden truth about human nature and more like a convenient fiction. Real people do not live as calculating machines. They hesitate, overreact, follow impulses, and sometimes choose the extra slice of pizza simply because it feels right. And although economic theory remains a powerful tool, it cannot capture the full texture of human life. Perhaps that is a blessing. A world populated entirely by cold calculators would be efficient, predictable, and profoundly dull, a little too neat for anyone's comfort.

As all this was unfolding, anthropologists were uncovering entire worlds in which the language of cost and benefit barely applied. In his classic work *The Gift*, Marcel Mauss showed how, in Polynesian societies, exchange was infused with spiritual force, or *mana*, and governed by elaborate systems of prestige, reciprocity, and obligation. Among the Kwakiutl of the Pacific Northwest, ritualised gift-giving ceremonies such as the potlatch involved extravagant, sometimes ruinous, displays of wealth. In many Native American traditions, including that of the Haudenosaunee Confederacy, decision-making is guided by the seventh-generation principle, which asks leaders to consider the consequences of their actions seven generations into the future. This moral horizon stretches far beyond modern profit cycles and exposes how narrow the rational actor model becomes when severed from culture, ethics, and time.

These acts were not designed to maximise individual gain but to establish social rank and strengthen communal bonds. They also reveal a common truth: human value, meaning, and responsibility cannot be fully captured by spreadsheets or algorithms. And set against such practices, the 'rational actor' model begins to look strikingly narrow, even parochial.

However, the ideal of economic rationality did not remain confined to economics. It escaped its disciplinary boundaries and steadily infiltrated institutions and everyday life. Today, cost-benefit analysis governs fields as varied as healthcare, education, and even romance. Hospitals rely on metrics such as quality-adjusted life years to decide who receives scarce organ transplants. Elite universities promote the net present value of their degrees. Dating apps deploy proprietary matching algorithms that treat intimacy as a problem of optimisation. Even environmental conservation has been drawn into this logic through ecosystem service valuation, which assigns monetary values to bees' pollination or to a coral reef's role in flood protection. In the early 2010s, Louisiana wetlands were often assigned per acre monetary values, sometimes ranging into the tens of thousands of dollars, in cost benefit analyses used to weigh development against preservation in planning models.

It is essential to note that this expanding economisation rests on a powerful assumption: that all goods can be translated into a single, commensurable currency, usually money or utility. The distortions introduced by this flattening are profound. How does one meaningfully price the grief of a lost language, or the imaginative future erased by the extinction of a single rainforest frog? Political philosopher Elizabeth Anderson has sharply criticised what she calls the 'tyranny of markets,' arguing that when moral and civic goods are forced into economic frameworks, they lose their very essence, much like trying to bottle a river.

Yet this reduction of value to calculation is not an eternal feature of Western thinking. Even within Western intellectual history, economics was not always dominated by cold rationality or the single-minded pursuit of profit. Thomas Aquinas, drawing on Aristotle, developed the idea of *justum pretium*, or the just price, which was grounded not only in supply and demand but in equity, moral responsibility, and the well-being of the community. Charging more than was fair, even when the market allowed it, was considered a moral failing. In seventeenth-century Amsterdam, merchant guilds often went further by regulating both wages and profits, placing limits on how much traders could earn in an effort to curb greed and maintain social stability. These concerns were sharpened by the shock of Tulip Mania in the 1630s, one of the earliest recorded speculative bubbles. At its height, a single rare tulip bulb could cost more than a skilled artisan's annual income, and in some cases more than a house in Amsterdam. When the market collapsed in 1637, fortunes evaporated overnight and contracts were rendered worthless. Notably, early forms of Dutch insurance were sometimes organised through church-run mutual aid societies, blending financial practice with ethical and religious life.

Even Adam Smith, so often reduced to the slogan of the 'invisible hand,' devoted far more attention to moral psychology than to markets alone. In *The Theory of Moral Sentiments*, he argued that sympathy, trust, and concern for others were essential to social and economic stability. Without them, markets would descend into exploitation and disorder. Smith's interest in Indian philosophy and admiration for aspects of Chinese meritocratic governance further complicate the image of him as a simple apostle of laissez-faire economics.

In the final analysis, modern economics excels at mapping efficiency and optimisation, but it often compresses human life into smooth curves and elegant abstractions. Beneath those equations stands a far more complex being: generous and grasping, rational and impulsive, guided not only by prices and incentives but by stories, values, loyalties, and dreams. Taken together, these tensions suggest that the world we inhabit is richer, stranger, and more relational than economic rationality alone can explain.

Breaking the Chains of Objective Determinism

As we have seen, in the early period, the universe was imagined as a clockwork automaton. Early physicists conceived it as a vast, precisely ticking mechanism. If you knew the position and speed of every gear, every planet, every particle and every force, you could, in principle, predict everything that followed. Tomorrow's weather, next year's harvest, even your next sneeze would unfold with mechanical certainty. This was the promise of classical physics: a world that was orderly, predictable, and governed by strict chains of cause and effect.

The twentieth century, however, introduced a quiet but profound rupture. Experiments at the smallest scales began to reveal behaviour that no clockwork model could accommodate. Particles did not act like tiny billiard balls moving along fixed paths. Instead, they hovered in states of possibility. A photon, for instance, could pass through two paths at once until an observation forced a decision. This is superposition, not like a coin flipping in the air, but like a coin that is both heads and tails at the same time.

Then came entanglement, the strangest feature of quantum theory. When two particles interact, they can become linked in such a way that measuring one immediately reveals information about the other, whether it is a metre away or a light-year distant. The outcomes are correlated beyond anything classical physics can explain. Crucially, no signal travels between them. There is no transmission of information, no hidden push or pull. It is not action at a distance, but connection without communication.

Einstein was deeply unsettled by this, dismissing it as 'spooky.' Abner Shimony was even more dramatic, and even poetic, calling it 'passion at a distance.' Yet Bell's theorem, followed by decades of increasingly precise experiments, confirmed the phenomenon. Nature, it turns out, is more deeply interconnected than classical thinking allowed, though not through familiar causal chains.

The unsettling implications do not stop there. Physicists such as Niels Bohr argued that quantum properties do not even exist in a definite form until they are measured. It is not merely that we lack knowledge. Reality itself may remain undecided until a question is posed. Heisenberg sharpened this insight further by showing that precision has limits built into nature. The more accurately a particle's position is known, the less precisely its momentum can be determined. This is not a flaw in our instruments but a feature of the universe itself.

The result is a profound shift in perspective. The universe is not a machine that simply runs on fixed rules, indifferent to observation. It is something more enigmatic, coherent yet saturated with ambiguity. Knowledge and reality are intertwined, not through simple cause and effect, but through a delicate and deeply strange interdependence.

Before we observe it, the quantum world sways in the shadows of possibility. At this level, certainty gives way to probabilities. Tiny particles—electrons, photons, etc.—do not have fixed properties until they are measured. They can exist in multiple states at once, as if undecided. Erwin Schrödinger illustrated this strangeness with his famous thought experiment: a cat inside a sealed box, linked to a quantum event, could be considered both alive and dead until the box is opened. This example highlighted the counterintuitive logic of quantum mechanics.

It is also worth noting that quantum theory does not overthrow classical physics; it expands it. For everyday objects, classical laws still hold. But at the tiniest scales, reality behaves more like a web of probabilities than a chain of causes. And at the opposite end of the scale, complexity theory reshapes how we understand large, interconnected systems. Whether it is the climate, a forest, or an economy, these systems do not behave in linear, predictable ways. You cannot understand a flock of birds by studying one bird, nor predict a hurricane by measuring a single gust of wind. Such systems exhibit emergent properties, i.e., behaviours that arise from the whole, not from the parts alone.

Ilya Prigogine, the Belgian physicist, extended this insight by studying systems that resist settling into equilibrium. From whirlpools to living cells, these dissipative structures grow and evolve by staying far from balance. Rather than breaking down in the face of change, they use it to build new forms of order—a discovery that earned him the Nobel Prize. Altogether, these ideas challenge the notion of the universe as a closed, deterministic machine. Reality, whether at the smallest or largest scales, is dynamic, relational, and full of surprises. To understand it, we must think less like watchmakers and more like ecologists, dancers, or poets.

This perspective resonates with other critiques of universalist science, including the work of Vandana Shiva, the Indian environmental activist and scholar. Shiva critiques the global dominance of Western science and its ecological consequences, emphasising that knowledge should not be homogenised or reduced to objective, universal facts. Her advocacy for biodiversity and indigenous knowledge champions the importance of respecting diverse systems of understanding, calling for 'earth democracy'—a vision that promotes harmony between human society and the earth. She also warns against the absolutist framing of initiatives like the so-called green economy:

> 'Squeezing the lives of people is now being proposed as the saviour of the planet. Through the green economy, an attempt is being made to technologize, financialise, privatise, and commodify all of the earth's resources and living processes.'

Shiva's critique reminds us that knowledge and action are inseparable from context, culture, and ethics. This is an essential counterpoint to the overly mechanistic or universalist approaches that dominate modern science and economics.

The renowned Indian philosopher I. K. Taimni similarly criticised the Western scientific worldview for ignoring the spiritual and metaphysical dimensions essential to understanding reality. Taimni argued that true knowledge cannot be confined to materialistic, objective frameworks; it also requires deep, subjective engagement with the cosmos, grounded in the experiential traditions of Vedanta and yogic philosophy. His perspective highlights how traditional Indian thought has long emphasised the unity of subjectivity and objectivity—an approach often missing in Western reductionist paradigms. Notably, this unity is not just an abstract concept in Puranic stories; it is woven into the very fabric of life.

Alternative Epistemologies

The scientific method has launched satellites, sequenced genomes, and brewed your instant coffee. But stepping into alternative epistemologies can feel like wandering into a vibrant desi bazaar of ideas, where logic meets lore and formulas dance with folk tales. This is not about ditching the lab coat; it is about wearing it alongside a rudraksha mala.

In a world that often feels like it is having an existential meltdown—climate change, rising seas, and AI-induced anxiety—what if we stopped relying on a single way of knowing? What if we paired scientific precision with spiritual depth, empirical rigour with intuitive insight? Take the chakra. It may not appear in a physics textbook anytime soon, but it quietly reminds us to stay balanced, aligned, and human. Western science has its undeniable gifts. Yet when you dip your toes into Indian knowledge systems, something shifts. It is like switching from grayscale to technicolour. The forest, then, is no longer merely a resource; it becomes a grandmother with her treasure chest of tales. The body is not a machine; it is a sacred temple. The cosmos is no longer cold and indifferent, but alive. These traditions do not only ask what is; they ask what matters. Perhaps wisdom was never meant to be a single-flavour snack. Maybe it is time to build a thali of knowing—scientific and sacred, spicy and sweet—and feast our way into a future that is not just smarter, but wiser.

American pragmatism, developed by thinkers like William James and John Dewey, offers a parallel challenge to rigid notions of truth. Instead of chasing a timeless, universal reality, pragmatism asks a simpler question: what works? Think of making chai. There is no single correct recipe. The blend changes with place, weather, and mood. Sometimes it is the ginger, sometimes the cardamom, that makes it just right. Pragmatism embraces this flexibility. It values ideas not for their abstract perfection, but for their ability to help us navigate life's uncertainties. Central to this view is fallibilism: the recognition that knowledge is always provisional, always open to revision. What works today may need rethinking tomorrow, and that is not a failure. It is how understanding grows. Truth, for pragmatists, lies not in being final, but in being useful, adaptable, and responsive to experience.

A similar insight emerges from our old friend: quantum physics. Niels Bohr's principle of complementarity suggests that reality cannot be captured from a single viewpoint. At the quantum level, particles can behave as waves or as particles, depending on how we observe them. These descriptions contradict one another, yet both are necessary. Bohr's lesson was simple but radical: to understand the whole, we must hold multiple perspectives at once. Perhaps truth is not a solitary monument, but a shifting landscape. A little of this, a little of that, shaped by where we stand and how we choose to look.

What if we applied this way of thinking to everyday life? Instead of chasing the elusive idea of a single "ultimate truth," we might accept that the world is simply too complex to be captured from one angle alone. Take a living organism. You can study it as a biological entity, focusing on its biochemistry, DNA, and cellular processes. You can also understand it as a historical being, shaped by evolution and time. Or you can examine it as a social creature, defined by its behaviour and relationships. Each perspective reveals something real, yet none of them can claim to be the whole truth.

This is not unlike the ancient Indian idea of Brahman, an ultimate reality that transcends all forms while containing them. It is not one thing among others, but the ground on which all perspectives coexist. Why reduce reality to a single lens when multiple views can illuminate different facets of the same whole?

Here is where things become interesting. It requires letting go of our urge to control every detail. For centuries, we have pursued certainty as if it were the Holy Grail, convinced that finding the "right" answer would finally bring order to the universe. That mindset now feels as dated as dial-up internet. Across disciplines, quantum physicists, evolutionary biologists, and philosophers are recognising that uncertainty is not a flaw in our understanding. It is the engine of creativity, innovation, and discovery.

Heisenberg's Uncertainty Principle makes this point with striking clarity. The more precisely you measure one property of a system, the less precisely you can measure another. Reality resists complete capture. The universe, it seems, thrives on a degree of mystery. Rather than fearing this, we might see it as an open invitation to think differently and imagine anew.

This insight is hardly new. The *Srimad Bhagavad Gita* urges us to act without attachment to outcomes, offering an ancient lesson in living with uncertainty. In a world that often feels chaotic and unpredictable, this shift in perspective can feel like a cheat code for navigating life. Uncertainty becomes fertile ground, not a void. New ideas take root there.

Knowledge does not grow in tidy boxes. It is messy, organic, and constantly evolving, much like the layered complexity of spices in a good curry. As both scientists and philosophers remind us, the capacity to adapt is far more powerful than the illusion of absolute control. And perhaps that adaptability is where real wisdom, and a touch of magic, quietly emerges.

Living beyond the Illusion

The Western fixation with objectivity, straight lines, and clear-cut answers has taken us a long way, from Moon landings to smartphones that can tell whether you are walking, running, or simply having a bad day. The achievements are undeniable. Yet when we step back and confront today's defining challenges, such as climate change, disappearing species, widening inequality, and cultural conflict, that clean and efficient logic begins to lose some of its shine. These are not problems that yield to a single equation or a master algorithm.

They are closer to Indian ragas: layered, evolving, and sensitive to nuance, where every note, pause, and rhythm shapes the whole. Miss one element and the entire mood shifts. Global issues arrive entangled in feedback loops, competing values, and deep uncertainty. Embracing this complexity does not mean abandoning reason or truth. It simply means acknowledging that life rarely unfolds as a straight line from point A to point B. It resembles a crowded bazaar instead, where the path changes depending on who you encounter, what draws your attention, or whether it is time to stop and talk.

This is not a rejection of science, but an invitation to nuance. It asks us to loosen our grip on rigid categories and accept that truth may appear differently from different vantage points. Context matters. Needs differ. Ways of knowing multiply. That shift in attitude may well be the beginning of wisdom.

Imagine what this could look like in practice. In schools, we might teach frameworks that encourage flexibility rather than reward only the 'correct' final answer. In governance, we could listen to a wider range of voices alongside experts, ensuring that policy is shaped not only by technical knowledge but also by lived experience. In economics, we might design systems that recognise human well being in its many forms, rather than measuring progress solely by GDP. In culture, rather than dismissing what feels unfamiliar, we could approach it with curiosity, much as we do when encountering a new musical tradition.

The comfort of absolutes can be reassuring, like knowing the rules of a game before it begins. But that comfort fades when the game itself keeps changing. Letting go of the demand for total certainty allows us to respond to change with greater resilience. Recognising that no single perspective, including our own, holds all the answers opens the door to a more honest and engaged relationship with both people and the planet.

Moving beyond absolutes does not lead to confusion or chaos. It leads to a richer understanding of our place in a world too vast to be captured by one truth alone, yet always open to those willing to approach it with humility and an open mind.

> 'If quantum mechanics hasn't profoundly shocked you, you haven't understood it yet.'
>
> — **Niels Bohr**

chapter 2

the quantum rabbit hole

In the summer of 1927, a group of physicists gathered in Brussels for what would become known as the Fifth Solvay Conference, one of the most pivotal meetings in the history of quantum mechanics. In the grand ballroom of the Hotel Metropole, intellectual energy crackled as twenty-nine scientists, including seventeen future Nobel Prize winners, assembled to debate nothing less than the nature of reality itself. Outside, Brussels moved along as usual, unaware that inside those walls humanity's understanding of the universe was on the verge of a profound shift. What was about to unfold was not merely a technical disagreement, but a clash of worldviews.

Einstein emerged as the clearest representative of one side of that divide, the most famous scientist in the world, standing firmly for a universe governed by classical, deterministic laws. It was a clockwork universe in which causes led cleanly to effects, particles possessed definite positions and velocities, and reality existed independently of observation. His universe resembled a cathedral of reason, illuminated by the steady light of causality. It had curvature, yes, but little room for chaos. In this picture, if you knew the complete state of the universe at one moment and had sufficient mathematical power, you could, in principle, predict its future indefinitely.

Opposite him stood Niels Bohr, the Danish physicist whose revolutionary work on atomic structure had earned him the Nobel Prize five years earlier. Bohr championed the emerging framework of quantum mechanics, a view of reality that was probabilistic and uncertain, where particles could exist in multiple states at once and where the act of measurement helped shape what was observed. Quantum mechanics felt less like a cathedral and more like a jazz club. It was spontaneous, probabilistic, and often counterintuitive to the point of absurdity. In the quantum world, particles did not simply exist in well-defined states. They hovered in superpositions, flirted with multiple possibilities at once, and only settled on a definite outcome when measured.

At the heart of their disagreement lay a fundamental question about the nature of reality. Einstein maintained that there existed a complete, objective world independent of our measurements, a reality that persisted whether or not we observed it. Bohr, by contrast, argued that physics could speak meaningfully only about what can be observed and communicated. The issue was no longer simply what the electron is doing, but what we can legitimately say about it. This was not merely a technical dispute but a deeper clash over the nature of knowledge itself, with profound metaphysical stakes concealed beneath the mathematics.

What made this moment truly historic was not only the clash between two extraordinary minds, but what that clash symbolised. At stake was a confrontation between absolutism and non-absolutism in scientific thought. The absolutist position held that nature followed precise, deterministic laws and that objects possessed definite properties regardless of whether they were observed. This view had shaped Western science since Newton. Laplace pushed it to its logical extreme, arguing that an intelligence that knew the position and force acting on every atom at a single moment could calculate both the future and the past with complete certainty.

Such thinking aligned comfortably with everyday experience. We do not see baseballs occupying multiple locations at once, nor do we expect coffee cups to tunnel through tables, however thin the barrier beneath them. Yet at the quantum scale, the domain of atoms and subatomic particles, this reassuring certainty collapsed. Experiments repeatedly confirmed the unsettling predictions of quantum mechanics: electrons behaved like waves when unobserved and like particles when measured; properties remained undefined until observation occurred; entangled particles influenced one another across vast distances. These findings challenged not only classical physics, but our most basic intuitions about how reality ought to work. The universe, it seemed, was far stranger and far less obedient to common sense than anyone in that ballroom had once imagined.

In the end, neither Einstein nor Bohr claimed a definitive victory. Their final intellectual battleground became the question of separability, whether physical systems located in different regions of space could be said to possess independent reality. Einstein articulated this concern most forcefully in a March 1948 letter to Max Born, where the debate reached its most philosophical pitch. Reflecting on what it means to speak of physical reality, he argued that whatever we call 'real' must, in some meaningful sense, be localised in space and time. What exists in one region of space, he insisted, should not depend on measurements performed in another. If events in region B can change merely because of a measurement in region A, then either quantum theory is incomplete, or we must abandon the idea that spatially separated systems possess independent existence. Einstein found the latter deeply troubling, confessing that his physical instincts recoiled from such a conclusion. Yet without separability, he wondered, what exactly is physics describing?

Despite this final and inconclusive clash, we are still living with the consequences of their arguments. Modern technology is built on both sides of the divide. GPS systems function only because Einstein's relativity precisely describes how time bends under gravity and motion, while computers, lasers, and MRI machines exist because quantum theory captures the behaviour of the microscopic world with astonishing accuracy. Our technological civilisation runs on this uneasy partnership.

Therefore, contemporary physics presents a strange kind of poetry: a universe in which large-scale structures unfold with elegant predictability, while the smallest constituents shimmer with uncertainty. Causality still governs the cosmos, yet at sufficiently small scales, the future seems unwilling to settle until it is observed. Perhaps the universe does play dice. Or perhaps it is subtler than that. Either way, the game is serious.

The implications of this strangeness soon found their sharpest expression through Werner Heisenberg in 1927. His uncertainty principle, published just before the legendary Solvay Conference where Einstein, Schrödinger, and Bohr sparred over the nature of reality, shattered the dream of perfect precision in physics. With one stroke, it cracked Newton's glasshouse of certainty.

His principle made a startling claim: the more precisely you know a particle's position, the less precisely you can know its momentum, and vice versa. This limitation is not due to faulty instruments or clumsy observers. Even with the best possible measuring devices, the uncertainty remains. At the quantum level, reality itself does not offer both facts at once. Nature is not hiding the truth from us; there is no fully determinate truth to uncover until measurement occurs. It is as if the universe allows you to trace its footsteps or feel its rhythm, but never both at the same time.

Around this intellectual turning point entered Max Born, a mathematician whose contribution reshaped how prediction itself was understood. Born gave quantum mechanics its statistical foundation. Particles, he argued, do not follow fixed paths. Instead, they move within fields of probability, and all we can ever know is the likelihood of finding them in a particular place when a measurement is made. As Born put it, 'The motion of particles follows probability laws, but the probability itself propagates according to the law of causality.' The behaviour of the particle remains uncertain, but the unfolding of that uncertainty obeys a precise mathematical order.

Born was also careful to steer between extremes. Reflecting on belief and reason, he wrote:

> 'There are two objectionable types of believers: those who believe the far-fetched, and those who believe that "belief" must be discarded and replaced by the "scientific method." Between these two extremes, there is enough scope for believing the reasonable and reasoning on sound beliefs.'

Here's an interesting side note: Born was also thegrandfather of Olivia Newton-John. Yes, that Olivia who sang 'You're the One That I Want' in *Grease*. As it turns out, quantum physics is only one degree removed from a Broadway-meets-bubblegum musical.

Returning to the deeper significance of the uncertainty principle, its implications reach well beyond the realm of physics. In Indian philosophical traditions, the idea that reality depends on how it is approached is ancient. The Upanishads explored the relationship between observer and observed long before Heisenberg translated it into equations. The method of *neti neti* (not this, not that) insists that reality cannot be captured in totality. Try to grasp it completely, and it slips away. Heisenberg's principle echoes this insight: certainty collapses when you demand too much of it.

Seen this way, the uncertainty principle is not merely a technical result; it is a philosophical blow to centuries of scientific absolutism. It tells us that uncertainty is not a flaw to be eliminated but a fundamental feature of nature. Even with perfect instruments, certainty does not arrive, because the deeper truth is that reality itself does not conform to it.

Niels Bohr took this aspect seriously. Rather than resisting paradox, he embraced it as central to the structure of the quantum world. His answer to quantum strangeness was the principle of complementarity. Quantum objects such as electrons do not possess a single, fixed identity. Depending on the experimental setup, they can appear as particles or as waves. Each description is valid, but only within its own context. You never get both views at once, and this is not a failure of technology or imagination.

For Bohr, this was not a temporary limitation waiting to be overcome. It was the way reality works. Quantum phenomena do not come with preassigned properties independent of observation. What appears depends on how the question is asked. Reality is not either-or; it is both-and, though never simultaneously.

Bohr captured this insight succinctly:

> 'Evidence obtained under different experimental conditions cannot be comprehended within a single picture, but must be regarded as complementary in the sense that only the totality of the phenomena exhausts the possible information about the objects.'

If Bohr was right, then reality itself resisted any absolute description. No single framework, however sophisticated, could fully capture quantum behaviour. Different experimental setups seemed to draw out different aspects of reality, as though nature wore multiple costumes and revealed only one depending on how the question was asked. It was as if the universe were saying, 'I contain multitudes, but you encounter them one at a time.'

The universe did not leave this question to interpretation alone. It answered instead through experiment, most famously through the double-slit experiment, perhaps the most unsettling demonstration in all of quantum physics. Think of it as stage magic performed by particles, where the trick changes depending on whether the audience is watching closely or not.

In 1927, Clyde Davisson and Lester Germer showed that electrons diffract when scattered off a nickel crystal. Instead of behaving like tiny bullets, they produced interference patterns characteristic of waves. Later experiments by Claus Jönsson and Akira Tonomura deepened the mystery. When electrons were sent through two slits one at a time, they still built up the same interference pattern. Each electron behaved as though it had passed through both slits and interfered with itself. Then came the twist. When detectors were installed to determine which slit the electron actually passed through, the interference pattern disappeared. The electrons behaved like ordinary particles again. Observation did not merely reveal reality; it helped determine which version of reality appeared. The experiment suggested something radical: truth is not a static object waiting to be uncovered but a dynamic process that unfolds through interaction. The observer is not outside the drama but part of the cast.

This insight, still taught in physics classrooms today, revealed a startling fact. The observer cannot be cleanly separated from the system being observed. Measurement changes outcomes in ways that classical physics cannot explain. As physicist John Wheeler later put it:

> 'The past has no existence except as it is recorded in the present. . . . No phenomenon is a phenomenon until it is an observed phenomenon. The universe does not exist "out there" independent of all acts of observation. Instead, it is in some strange sense a participatory universe.'

By the early 1930s, quantum mechanics had moved from a radical proposal to a fully developed mathematical framework. With the Schrödinger equation, physicists could predict how quantum systems evolved, not as definite paths but as waves of probability. These were not physical waves like ripples on water. They were probability distributions, indicating where a particle might be found if a measurement were made. And in this new picture, particles do not settle into a single state on their own. They exist in superposition, a coexistence of all possible states at once. Only when a specific question is asked through measurement does one outcome appear, as if reality makes a choice at the moment it is queried.

To expose the strangeness of this idea, Erwin Schrödinger proposed a thought experiment in 1935 that has since become iconic. A cat is sealed in a box, its fate linked to a quantum event such as the decay of a radioactive atom. If the atom decays, the cat dies; if it does not, the cat lives. According to quantum theory, until the box is opened, the atom exists in a superposition of decayed and not decayed, and the cat's state remains unresolved.

Schrödinger did not mean this literally, nor did he suggest that real cats hover between life and death. His point was sharper. Quantum theory, while mathematically successful, does not explain when or how possibility becomes actuality. The cat exposes a gap between the equations and our understanding of reality. That is why Schrödinger's cat continues to stalk philosophy seminars, physics classrooms, and popular culture. It serves as a reminder that in the quantum world, reality does not fully take shape until attention is paid. Even existence, it seems, waits for a question.

Fielding the Revolution

In the early days of quantum mechanics, physicists were fixated on particles. Electrons, photons, and their kin took centre stage, puzzling scientists with behaviour that seemed erratic and unpredictable. Yet behind this focus on particles, a deeper and more radical shift was underway: the emergence of quantum field theory.

Before quantum field theory, fields such as the electromagnetic field were treated as passive, continuous backgrounds that simply filled space. Quantum field theory overturned this picture. Fields were no longer secondary to particles; they became the fundamental entities of nature. Particles themselves were reinterpreted as excitations of these fields, brief disturbances rippling through an underlying fabric. An electron, in this view, is not a tiny object moving through space but a localised vibration in the electron field. Reality, it turned out, was less like a collection of solid beads and more like a restless ocean, where what we call particles are momentary waves.

At the heart of this conceptual breakthrough stood Paul Dirac, a physicist determined to reconcile quantum mechanics with Einstein's theory of relativity. In pursuing mathematical consistency and elegance, Dirac did more than resolve technical problems. He predicted the existence of antimatter, a mirror form of matter identical in mass but opposite in electric charge. At first, the idea seemed fantastical. Then, in 1932, the positron, the antimatter counterpart of the electron, was experimentally discovered, confirming Dirac's bold insight. When matter and antimatter meet, they annihilate each other, releasing energy. This led to quantum fields being seen not just as abstract constructions, but as descriptions of a universe that was real, testable, and far stranger than anyone had anticipated.

In his quest, Dirac was guided not only by logic but by a deep faith in mathematical beauty and symmetry. As he famously observed:

> 'It seems to be one of the fundamental features of nature that fundamental physical laws are described in terms of a mathematical theory of great beauty and power, needing quite a high standard of mathematics for one to understand it . . . We simply have to accept it. One could perhaps describe the situation by saying that God is a mathematician of a very high order.'

The story continued in the 1940s, when Richard Feynman, Julian Schwinger, and Sin-Itiro Tomonaga developed quantum electrodynamics, the first fully successful quantum field theory. Quantum electrodynamics explained how light and matter interact with extraordinary precision. Feynman, in particular, transformed the way physicists thought about these interactions. He introduced what are now known as Feynman diagrams, visual tools that allowed complex particle processes to be represented as simple drawings.

These diagrams were more than calculational shortcuts. They reflected a deeper shift in perspective, one in which interactions, relationships, and processes took precedence over isolated objects. The universe, as revealed by quantum field theory, was not built from particles acting alone but from fields constantly interacting, exchanging energy, and giving rise to the fleeting phenomena we experience as matter and light.

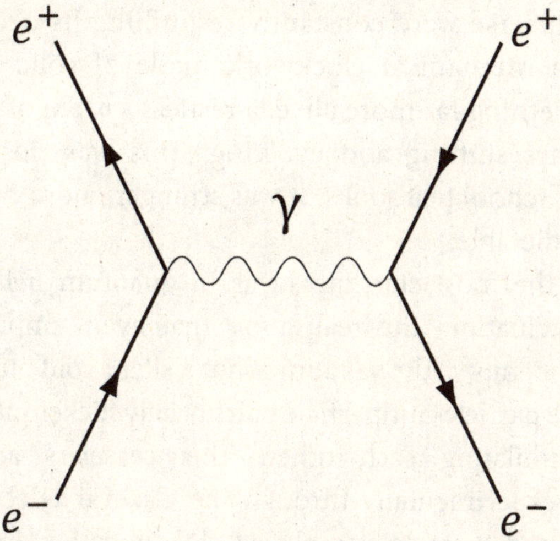

A Feynman Diagram of an electron and its antimatter positron meeting and annihilating each other, with the emission of a photon (energy), which in turn can give rise to a pair of electrons and positrons.

Instead of being solid, unchanging objects, particles came to be understood as dynamic ripples in a quantum sea. Nothing was fixed; everything was in flux. Richard Feynman embraced this non-absolutist spirit wholeheartedly, as his own words reveal:

> 'I think it's much more interesting to live not knowing than to have answers which might be wrong. I have approximate answers and possible beliefs and different degrees of uncertainty about different things, but I am not absolutely sure of anything . . . I don't feel frightened not knowing things, by being lost in a mysterious universe without any purpose, which is the way it really is as far as I can tell.'

With the rise of quantum field theory, reality itself began to feel fluid. Particles could appear and vanish in an instant, as if the universe were constantly reshuffling its deck. The old image of a mechanical clockwork made of solid parts gave way to something far more alive: a restless system of interacting fields, always shifting and evolving. This was no longer the universe of school textbooks. It was stranger, more creative, and far less predictable.

One of the most startling ideas in quantum field theory is vacuum fluctuation, the realisation that even 'empty' space is anything but empty. The vacuum is not a silent void but a seething arena where particle-antiparticle pairs briefly flicker into existence before annihilating each other. This ceaseless activity was confirmed experimentally through the Casimir effect, in which two uncharged metal plates placed close together in a vacuum attract one another due to pressure differences created by these virtual particles. This showed that empty space is dynamic and energetic, not inert.

These virtual particles are not merely mathematical curiosities. They play a central role in the fundamental forces of nature. The electromagnetic force, for instance, is not the result of direct contact between charged particles. It arises from the constant exchange of virtual photons that carry momentum between them. Without this invisible traffic, atoms could not form, matter would not hold together, and light itself would cease to exist.

Then there is renormalisation, one of the most counterintuitive yet essential features of quantum field theory. When physicists first performed calculations in QFT, they encountered infinities that seemed to threaten the entire framework. Rather than being fatal flaws, these infinities turned out to be signals of the theory's depth. Renormalisation provides a systematic way to absorb these infinities and extract meaningful, finite predictions. It ensures that the laws of physics remain consistent whether we examine nature at very small scales or larger ones. Far from being a technical patch, renormalisation is woven into the fabric of the theory. It reflects a universe whose rules remain coherent across different levels of description.

Taken together, these ideas mark a profound shift in how we understand reality. The universe is no longer seen as a collection of solid objects moving through empty space. It is a web of interacting fields, with particles appearing as temporary disturbances in an underlying continuum. This transformation is nothing short of revolutionary. And while quantum field theory may feel remote from everyday experience, it quietly underpins much of modern life. Semiconductors, lasers, medical imaging, and even GPS depend on principles rooted in this strange and subtle framework. The next time you check your phone, it is worth remembering that the unpredictable world of quantum fields is working tirelessly in the background, making it all possible.

The field revolution carried profound philosophical consequences, pushing back against the absolutist assumptions of classical physics. Reality no longer appeared fixed or fully determinate. Even the identities and number of particles were not constant but could fluctuate. The vacuum itself, once imagined as sheer emptiness, was revealed to be a restless sea of virtual particles constantly appearing and disappearing. Far from being nothingness, it was rich with latent possibility. This vision resonates strongly with ancient Indian philosophy, especially the Buddhist concept of śūnyatā, where emptiness is not a void but a dynamic, generative ground from which forms arise.

The triumph of quantum electrodynamics lay not only in its conceptual daring but in the astonishing precision of its predictions. The electron's magnetic moment, for example, matched theoretical calculations to more than ten decimal places. This level of agreement has often been compared to measuring the distance between New York and Los Angeles with the accuracy of a human hair's width. Such precision established QED as more than an elegant theory. It became a practical and extraordinarily reliable framework for understanding the subatomic world.

After this success, physicists were not content to stop. The 1960s and 1970s witnessed an ambitious attempt to go further towards unification. The result was the Standard Model, a quantum field theory that brought together three of the four known fundamental forces: electromagnetism, the strong force, and the weak force. It also organised the growing menagerie of elementary particles discovered in accelerators from Berkeley to Geneva. In effect, it became a kind of periodic table for quantum physics, cataloguing quarks, leptons, force-carrying bosons, and eventually the Higgs boson, whose existence was confirmed only in 2012.

Steven Weinberg, one of the architects of this unification, once remarked with dry humour, 'The Standard Model is so complex it would be hard to put it on a T-shirt, though not impossible; you would just have to write kind of small.' The joke carried a deeper truth. Physicists had deciphered much of nature's script, diagram by diagram, yet the script itself remained oddly incomplete.

Despite its remarkable success, the Standard Model came with a troubling limitation. It depended on nineteen parameters, including particle masses, coupling strengths, and mixing angles, that could not be derived from the theory itself. They had to be measured experimentally and inserted by hand. It was like composing a symphony while needing someone else to supply the notes for every instrument. The masses of particles, such as the electron or the top quark, were not explained. They simply existed.

There is an old story about scholars debating how many angels could dance on the head of a pin while Constantinople was under siege. Modern physics has its own version of this tension. Today, the angels are quantum fields, and the pin is the vast network of particle colliders built at enormous public expense. Even the Higgs boson, often nicknamed the 'God particle,' was not part of the original framework. The Higgs mechanism was introduced later to explain why particles have mass at all, giving the impression of a necessary but improvised addition rather than a natural consequence of deeper principles.

The result is a theory of immense power and undeniable elegance, yet one that feels provisional. The Standard Model explains neutrino oscillations, CP violation, and the stability of stars like the sun. At the same time, it leaves gravity entirely outside its scope. Dark matter and dark energy, which together make up about ninety-five per cent of the universe, remain unexplained. In this sense, the model quietly echoes insights associated with thinkers like Nāgārjuna or Gödel: no system can be fully complete on its own terms.

The Standard Model stands as one of humanity's greatest intellectual achievements. It is also a reminder of our limits. It works brilliantly, yet it points beyond itself, held together by empirical success and animated by the hope that deeper symmetries and more encompassing ideas still await discovery.

In 1983, the Standard Model was hitting its stride, and it scored a major victory with the discovery of the W and Z bosons, the heavy carriers of the weak nuclear force, whose measured masses matched theoretical predictions with uncanny precision. The real showstopper, however, arrived in 2012, when the Higgs boson, long theorised but never observed, finally emerged from the data at CERN. This was not merely another box ticked. It was experimental confirmation of the Higgs field itself, the invisible medium that gives particles their mass. Yet even amid celebration, the limitations of the Standard Model became impossible to ignore. Gravity was nowhere to be found. Dark matter and dark energy were not even mentioned, despite accounting for most of the cosmos. Neutrinos, those famously elusive particles, added further trouble by revealing that they do have mass, directly contradicting the model's original assumptions.

These gaps are not just technical inconveniences; they are philosophically significant. They remind us that the Standard Model, for all its elegance and predictive power, is not a final theory. It is better understood as a meticulously craftedprovisional scaffolding that is effective and beautiful but incomplete.

Brian Greene captures this sense of unfinished longing through Einstein's lifelong pursuit of a unified field theory:

> 'Seventy-five years ago this month, *The New York Times* reported that Albert Einstein had completed his unified field theory—a theory that promised to stitch all of nature's forces into a single, tightly woven mathematical tapestry. But as had happened before and would happen again, closer scrutiny revealed flaws that sent Einstein back to the drawing board. Nevertheless, Einstein's belief that he would one day complete the unified theory rarely faltered. Even on his deathbed, he scribbled equations in the desperate but fading hope that the theory would finally materialise. It didn't.'

This attitude resonates with much older perspectives, including Indian philosophical traditions such as Nyāya and Sāṃkhya, which treat knowledge as contextual, provisional, and continually refined. No model, however successful, is absolute.

In conclusion, like quantum fields themselves, our understanding must remain fluid and responsive, open to revision as new phenomena emerge. The Standard Model may be the Constantinople of modern physics, a gleaming capital of symmetry and structure, but it too awaits its renaissance, or perhaps its unification with the next great empire of thought. Until that deeper theory arrives, physicists continue navigating the unknown not with certainty, but with curiosity and humility.

The Stringscape

Once upon a spacetime, in the late 1960s, a bold new idea took root in the minds of physicists: what if the fundamental building blocks of reality were not tiny, point-like particles, but minute, vibrating strings? Yes, strings. Not the kind found on a violin or a kite, but unimaginably small filaments of energy, oscillating in dimensions far beyond the reach of our senses.

These were not mere flights of mathematical fancy. String theory offered a daring response to physics' most stubborn problem: gravity. For decades, gravity had been the odd one out. While electromagnetism and the nuclear forces fit comfortably within quantum mechanics, gravity stubbornly resisted quantisation. Calculations blew up into infinities, and the mathematics simply broke down. String theory changed the terms of the problem. It did not merely make room for gravity, but it also built gravity into its very structure, smoothing out the wild infinities that had long plagued physicists. Even more striking was what string theory implied about matter itself. The familiar zoo of particles—electrons, quarks, neutrinos, and the rest—was no longer fundamentally distinct. They were different vibrational modes of the same underlying entity. Each particle was a string playing its own note. In this view, the universe was made not of discrete things, but of resonances, a kind of cosmic music.

This was intoxicating. For the first time, it seemed genuinely possible to unite all the forces of nature—gravity, electromagnetism, and the nuclear forces—within a single elegant framework. Einstein had chased such a vision in his final years. As he once put it:

> 'I want to know how God created this world. I am not interested in this or that phenomenon, in the spectrum of this or that element. I want to know His thoughts; the rest are just details.'

For a moment, string theory appeared to whisper those thoughts. But nature, as ever, demanded a price. For the mathematics to work, string theory requires more than the three dimensions of space we inhabit. It needed ten, and in some versions, eleven. The extra dimensions were not missing; they were hidden, tightly compactified and curled up at every point in space, like elaborate origami folded beyond visibility. Strange as this sounded, it carried a deeper implication. Space and time themselves might not be fundamental. Reality could be woven from a structure far more intricate than everyday experience suggests.

The surprises did not end there. In the 1990s, physicists realised that there was not one string theory, but five, each internally consistent. It was as though five different maps led to the same concealed city. Rather than a failure, this multiplicity revealed something deeper: the five theories were simply different perspectives on a single underlying framework, which came to be known as M-theory. No one quite agrees on what the 'M' stands for. Mother and mystery are popular guesses.

Just when physicists thought string theory could not grow any stranger, it introduced a new character into the cosmic drama: the dilaton. You can think of the dilaton as a kind of cosmic volume knob. It is a scalar field that determines the size of the extra dimensions curled up within string theory's framework. But the real twist lies in what it implies. Depending on how the dilaton behaves, even the so-called fundamental constants of nature, numbers we treat as immutable, may not be constant at all.

Let that sink in. Quantities once carved into the tablets of physics as eternal truths might, under the right cosmic conditions, shift across space and time. These changes would not be arbitrary but would arise from the subtle interplay between extra dimensions and quantum fields. Nature's rulebook, it seemed, came with footnotes and fine print.

The surprises did not end there. From the study of black holes emerged another radical idea: the holographic principle. Developed by Gerard 't Hooft and Leonard Susskind, it suggested something deeply counterintuitive. Everything contained within a region of space—all its matter, energy, and information—might be fully encoded on that region's boundary. In effect, a three-dimensional universe could be a projection of information stored on a two-dimensional surface.

This was not science fiction, but a serious proposal rooted in quantum gravity. The image was unsettling yet elegant: reality as a cosmic hologram, a vast three-dimensional display generated from an unseen screen. Strangely, the idea echoed much older intuitions. In Indian metaphysics, the concept of māyā describes the world as a kind of appearance or projection that conceals a deeper reality. The holographic principle felt like physics arriving, by a very different route, at a similar insight—this time expressed in equations rather than sutras. And still, the rabbit hole went deeper.

As physicists pushed further into the mathematics of string theory, they encountered what came to be called the cosmic landscape. This was not a landscape of hills and valleys, but of possible vacuum states. There were not dozens or thousands of them, but potentially 10^{500}. Each vacuum corresponded to a different possible universe, with its own laws of physics, particle properties, and constants. Most of these universes would be lifeless or fleeting. But a tiny fraction might possess the delicate balance required for stars, planets, chemistry, and life. This perspective brought renewed attention to the anthropic principle. Perhaps the universe is not fine-tuned for us. Perhaps we exist because we happen to inhabit one of the rare universes in which beings like us can arise. The cosmos is not designed with us in mind; we are simply the result of surviving the cosmic lottery.

It is a dizzying conclusion, but also an oddly liberating one. String theory, for all its mathematical rigour, does more than extend physics. It forces us to rethink what we mean by reality itself. The universe begins to look less like a fixed stage governed by rigid rules and more like a shimmering, evolving symphony, where constants can drift, dimensions can curl, and what we take to be solid reality may be the most elaborate illusion of all.

The revelations of string theory did not merely adjust the dials of physics; they replaced the entire control panel. For centuries, the laws of nature were treated as immutable, etched into reality with the certainty of Newton's *Principia*. String theory unsettles that confidence. It asks a disquieting question: what if the constants we call 'universal,' such as the speed of light or the charge of the electron, are not universal at all? What if they resemble local dialects in a larger multiversal grammar, varying subtly across different regions of existence? This is not speculative fantasy. It is the natural consequence of allowing strings to vibrate in ten or eleven dimensions and of taking fields like the dilaton seriously. These fields suggest that the shape and size of the hidden, compact dimensions can influence quantities we once thought sacrosanct. Physicists often model these extra dimensions as Calabi-Yau manifolds, intricate curled geometries folded into every point of space. Their structure may quietly determine the strengths of forces, the masses of particles, and even the value of the Higgs field, gravity, electromagnetism, and particle properties, all become contingent on geometry.

As highlighted before, this openness resonates unexpectedly with the Jain doctrine of *anekāntavāda*, the idea that reality is inherently manifold and cannot be captured from a single standpoint. Like a gemstone that flashes different colours when viewed from different angles, truth emerges through perspective and context. This is not merely poetic; it is a deeply epistemological claim. Reality is not a single story told from one vantage point, but an ensemble performance in which meaning arises from relations and conditions. String theory makes a similar claim, though in the language of Planck scales rather than parables. Its most unsettling implication may be philosophical rather than technical. It pushes us towards a renewed humility about what we can know and how firmly we can know it. Long before modern physics, Indian thinkers from Nāgārjuna to the Advaitins argued that reality is not monolithic but relational, contingent, and elusive.

In this sense, string theory's true gift is not only mathematical but metaphysical. It invites us to see the universe less as a machine and more as a rāga, governed by structure, yet fluid, adaptive, and exquisitely sensitive to context. Perhaps its most radical suggestion is not that it might unify gravity and quantum mechanics, but that it quietly asks us to rethink what we mean by 'real.' And in a cosmos where laws can flex, dimensions can hide, and constants can drift, the only absolute may be the absence of absolutes. In that realisation, strangely and beautifully, we find not uncertainty, but a deeper kind of coherence.

The Cloud of Possibilities

Earlier, we saw how quantum mechanics thrives on paradox. It predicts the behaviour of particles with astonishing precision, down to decimal places that leave little room for doubt, yet it conceals a deep riddle at its core: the measurement problem. This is the philosophical sting in the tail, the ghost haunting the mathematics. How does a quantum system, perfectly content to exist as a superposition of possibilities, suddenly collapse into a single outcome the moment a measurement is made? Moreover, the equations themselves offer no clear answer. They describe a smooth and deterministic evolution of the wave function, right up to the point of measurement. Then the story abruptly changes. The wave function collapses, and what was once a spread of possibilities becomes a single, definite result. This is not a minor technical irritation but a genuine metaphysical puzzle.

To manage this discomfort, physicists in the 1920s developed what came to be known as the Copenhagen interpretation, shaped largely by Niels Bohr and Werner Heisenberg. Bohr, who moved easily between physics and philosophy, urged restraint. Do not ask what is 'really' happening, he advised. Calculate the probabilities, perform the experiment, and accept the outcome. In this view, quantum systems do not possess definite properties until they are measured. Before that, they exist in a state of indeterminacy, a cloud of potential rather than a catalogue of facts. But this move raised unsettling questions. What is so special about measurement? Why should the interaction with a detector, a screen, or an instrument compel nature to choose a definite state? This was not a small inconvenience. It threatened the classical picture of a universe unfolding independently of observation.

Both Bohr and Heisenberg accepted this challenge without hesitation. In many ways, they were early opponents of scientific absolutism. Heisenberg, in particular, was influenced by broader currents in German thought, including neo-Kantian philosophy, Goethe's reflections on perception, and the existential anxieties of his time. He expressed this position starkly when he wrote:

> 'I believe that the existence of the classical "path" can be pregnantly formulated as follows: The "path" comes into existence only when we observe it.'

This was not a mere poetic flourish. It was a direct rejection of the Newtonian idea that nature carries a complete, observer-independent narrative. Heisenberg suggested that physical reality remains unfinished until observation takes place. Nature, in this sense, does not simply reveal itself; it responds.

The implications of this approach were profound. Particles no longer possessed definite positions, velocities, or trajectories in and of themselves. In their place appeared uncertainty, entanglement, and a subtle interplay between observer and observed. This also implied that knowledge was no longer a passive mirror of reality but an active participant in its formation. And yet, for all its pragmatic success, the Copenhagen interpretation raised a new set of questions. Why does measurement cause collapse at all? What qualifies as a measurement? Is it a conscious observer, a mechanical device, or any interaction that leaves a trace? Why does the universe appear to wait for observation before settling on a definite outcome? These questions are not marginal curiosities. They strike at the foundations of what we mean by reality. Modern experiments involving delayed choices and entangled particles have only intensified the puzzle. Increasingly, quantum mechanics seems to suggest that reality is not merely uncertain but fundamentally relational. And in acknowledging this, Bohr, Heisenberg, and their contemporaries took an extraordinary step. They admitted that reality may not be something we merely discover. It may be something we help bring into being.

By the midtwentieth century, the Copenhagen interpretation had settled into polite orthodoxy. It was useful, mathematically elegant, and philosophically evasive, solving the equations while sidestepping deeper questions. These questions sat in lecture halls like an uninvited guest that everyone politely ignored.

In 1957, that silence was broken by an unlikely figure: Hugh Everett III, a graduate student at Princeton. With unusual confidence and formidable mathematical skill, Everett proposed what became known as the many-worlds interpretation (MWI). According to him, the wave function never collapses. Instead, every possible outcome of a quantum event occurs, and the universe branches accordingly. For example, when you measure an electron's spin, one version of you sees it point up while another sees it point down. The implications of this idea are dizzying. This means in principle, there is a universe for every coin toss, every decision not taken, and every fork in the road of personal history.

Unsurprisingly, the reaction to this claim was chilly. Everett's thesis was quietly shelved, dismissed as too extravagant and too close to science fiction. Yet the idea refused to disappear. Richard Feynman found it intriguing, and in the 1970s, Bryce DeWitt resurrected it with serious force. In his essay *Quantum Mechanics and Reality*, DeWitt wrote:

> 'All the worlds are there, even those in which everything goes wrong and all the statistical laws break down . . . We can perhaps argue that in those branches in which the universe makes a habit of misbehaving in this way, life fails to evolve; so no intelligent automata are around to be amazed by it.'

Over time, Everett's proposal moved from heresy to headline. As quantum computing gained momentum and the language of the multiverse seeped into both popular culture and serious mathematics, many-worlds began to look less like fantasy and more like a viable way to take quantum theory at face value. On the other side of the interpretational spectrum stood another dissenter: David Bohm. A student of Oppenheimer and later an intellectual exile due to McCarthyera politics, Bohm rejected both Copenhagen's ambiguity and Everett's proliferating universes. Instead, he returned to the unfashionable idea of determinism equipped with modern mathematics.

In Bohmian mechanics, particles are real and always have definite positions. Their motion, however, is guided by a pilot wave that evolves according to the Schrödinger equation. The theory reproduces all the predictions of standard quantum mechanics, but at a price. It is explicitly non-local. Particles can influence one another instantaneously, regardless of distance. This meant that what Einstein once mocked as 'spooky action at a distance' became central rather than accidental. And for Bohm, this feature was not merely a technical oddity, but it hinted at a deeper unity beneath apparent separateness. In his later years, this intuition drew him towards Indian philosophical ideas and into dialogue with J. Krishnamurti, with whom he explored themes of wholeness, consciousness, and the illusion of fragmentation.

Thus, we can see how Everett and Bohm took very different paths. One multiplied worlds and the other revealed hidden order. Yet both rejected the idea that reality waits passively for observation before it exists. Each, in his own way, refused the quantum shrug.

Just when it seemed the quantum world had revealed all its surprises, a new perspective made a spectacular entry: QBism, short for Quantum Bayesianism. It arrived less like a new theory of particles and more like a philosophical koan expressed in probabilities. If many-worlds gave us branching universes and Bohmian mechanics gave us invisible pilot waves, QBism did something even more radical. It questioned the very idea of an objective quantum reality.

According to QBism, the wave function does not describe something 'out there' in the universe. It represents an agent's personal expectations about future experiences. A quantum state is not a God's eye description of reality but a carefully updated ledger of beliefs, written in Bayesian ink. So, when a measurement is made and the wave function appears to collapse, QBism insists that nothing physical has collapsed at all. What has changed is the observer. Expectations are revised, much like reassessing a poker hand after an unexpected raise. Schrödinger's cat, in this view, is not a creature suspended between life and death but a vivid metaphor for fluctuating degrees of confidence.

Developed primarily by Christopher Fuchs and collaborators, QBism often sounds closer to a philosophy seminar than a physics lecture. Yet it is as mathematically consistent as any competing interpretation. Fuchs has described physics as 'a dynamic interplay between storytelling and equation writing,' arguing that neither stands alone, even in principle. He goes further, questioning the absoluteness of physical law itself:

> 'One way to look at it is that the laws of physics aren't about the stuff "out there." Rather, they are our best expressions of what our own limitations are . . . Methodologically we should assume that the universe is before us so that we can shape it, that it can be changed, and that it will push back on us. We will understand our limits by noticing how much it pushes back.'

What QBism challenges is not the predictive success of quantum mechanics but a deeply ingrained instinct: the belief that reality must exist independently of the observer, like a museum exhibit sealed behind glass. QBism breaks that glass and says that the universe does not simply wait to be discovered; it responds only when questioned, and the act of questioning matters. In this sense, QBism is an unapologetically anti-absolutist stance. Here, the situation takes an interesting turn. Everett's multiverses, Bohm's hidden variables, and QBism's belief centred account are empirically indistinguishable. They produce the same predictions, survive the same experiments, and earn the same Nobel Prizes. The physics is firm, but the metaphysics is wildly unsettled. This highlights how reality can wear different masks, depending on how we choose to interpret the mathematics. Perhaps that is why David Mermin's weary slogan, 'Shut up and calculate,' became so popular among physicists. Yet the persistence of these debates suggests they are not mere philosophical indulgences. The universe may be offering a subtler lesson: the more precisely we try to pin it down, the more it slips away. We then begin to confront an uncomfortable truth that the observer was never outside the puzzle. Instead, we have always been part of the riddle itself.

Few physicists captured this unsettling idea better than John Wheeler, who proposed the notion of a participatory universe. For Wheeler, the cosmos was not a cold machine running independently of us. In some profound sense, it may require observers not merely to watch the story unfold, but to help write it. And as strange as this sounds, this idea has an uncanny resonance with the ancient Indian philosophy of *Pratyabhijñā*, central to Kashmir Shaivism. In this tradition, consciousness is not a passive mirror reflecting an external world, but the very source through which the world appears at all. Stated differently, reality unfolds through an act of recognition, a continuous interplay in which awareness and existence cannot be separated. This suggests that you are not merely in the universe and that, in a deep sense, the universe is also in you.

Today, as quantum technologies advance and foundational questions resurface with renewed urgency, their oncemarginal ideas are being revisited with serious intent. What was long considered fringe now looks increasingly like the frontier. It turns out that the deepest questions about reality were never settled but were only postponed.

Quantum Non-Absolutism in the Laboratory

In the quiet laboratories of Vienna, something extraordinary took place. There was no cinematic flash, no single dramatic moment, but a series of exquisitely precise experiments that quietly reshaped our understanding of reality. Anton Zeilinger, a physicist with the curiosity of a philosopher and the discipline of an engineer, demonstrated that entangled particles, once linked, remain connected across vast distances in ways that defy classical ideas of cause and effect.

Throughout the 1990s and early 2000s, Zeilinger's team repeatedly confirmed what once sounded like science fiction. Measure one particle in an entangled pair, and the other reflects the outcome instantly, even if separated by mountains or continents. Einstein famously dismissed this notion, uneasy with the suggestion that nature might be violating his most sacred limit: the speed of light. But Zeilinger was not chasing ghosts. He was exposing something deeper, including our long-held desire to explain nature completely in terms of hidden causes. As he put it:

> 'We have tried for centuries to look deeper and deeper into finding causes and explanations, and suddenly, when we go to the very depths, to the behaviour of individual particles of individual quanta, we find that this search for a cause comes to an end. There is no cause. In my eyes, this fundamental indeterminateness of the universe has not really been integrated into our worldview yet.'

What Zeilinger's experiments revealed is that no signal or influence actually travels between the particles. Instead, the entangled pair behaves as a single, indivisible system, regardless of the distance between its components. It is as if two characters in a novel appear in different chapters but remain bound by the same narrative logic and the physical separation turns out to be secondary. What this means is that at the quantum level, connection is written into the structure of reality itself.

This phenomenon, known as non-locality, is not a minor technical curiosity. It is a direct challenge to the classical picture of the world as a chain of neatly ordered causes and effects. This shattered the final footholds of Newton's universe, where reality resembled a finely tuned watch, each gear driving the next. In Zeilinger's universe, reality resembled a jazz ensemble without a conductor—responsive rather than predetermined and not confined to local interactions.

In this light, quantum mechanics has little patience for absolutist worldviews. It does not merely suggest that reality is strange but shows that the theory itself cannot assign definite properties to systems before a question is asked. What 'is' depends on how we inquire. Thus, the universe begins to resemble a Rorschach test, revealing different answers depending on the frame we bring to it.

Philosopher David Pearce pushes this line of thought even further, beyond contextuality, towards a radical rethinking of matter itself. He argues that what gives life to the equations of quantum mechanics is not inert substance but subjectivity at its most fundamental level:

> 'I argue that what breathes fire into the QM equations is field-theoretic what-it's-likeness: "microqualia" to use a philosopher's term of art. The different values of the solutions to the ultimate physical equations exhaustively yield the abundance of different values of subjectivity. . . There is no "explanatory gap" because there aren't any mater**ial objects, not even brains or nerve cells as commonly (mis) perceived. Instead, over millions of years, non-equilibrium thermodynamics and universal, (neo-) Darwinian principles of natural selection have contrived to organise a minimal and self-intimating subjective sludge of microqualia into complex functional living units. Initially, these units have taken the form of self-replicating, information-bearing biomolecular patterns. Eventually, selection-pressure has given rise to complex minds as well, albeit as just one part of the throwaway host vehicles by which our genes leave copies of themselves. Conscious mind, on this proposal, is a triumph of organisation: our egocentric virtual worlds are warm and gappy QM-coherent states of consciousness . . . Moreover

> the idea that what-it's-likeness is the fire in the equations is (at least) consistent with orthodox relativistic quantum field theory because the theorists' key notions (e.g. that of a field, string, brane, etc.) are defined purely mathematically. In other cases, they readily lend themselves to such a reconstruction. Using the word "physical" doesn't add anything of substance.'

This idea flips the script on traditional physics. Instead of solid material objects doing all the causal work, it is quantum fields themselves, buzzing with microqualia (a shorthand for tiny units of subjective experience), that drive the universe's equations. There is no dualism here, and no mysterious material objects like brains set apart from the rest of nature. There are only fields interacting in ways that give rise to both the external world and our inner experience. It is a genuinely mind-bending picture. Additionally, these quantum quirks do not live only in philosophical speculation. They power one of the most radical technologies on the horizon: quantum computing. If classical computers are glorified abacuses, calculating with bits that are either 0 or 1, quantum computers are something altogether stranger and far more superior. They use qubits, which can be 0 and 1 at the same time through superposition, and can also share eerie correlations through entanglement, allowing them to work together like psychic twins. Instead of checking one solution after another, they explore entire landscapes of possibilities, collapsing only when forced to do so, like Borges' library of infinite paths.

David Deutsch offered a characteristically Everettian take on this, writing:

> 'Quantum computation is a distinctively new way of harnessing nature . . . It will be the first technology that allows useful tasks to be performed in collaboration between parallel universes.'

Another point to note here is that what makes quantum computers so powerful is not merely speed. Their true advantage lies in doing what once seemed impossible: factoring enormous numbers that would take classical computers longer than the age of the universe.

The irony here is hard to miss. The deeper we go into quantum hardware, the clearer it becomes that we are building machines not out of certainty, but out of uncertainty itself. Quantum computing runs not on what is, but on what might be. It is like constructing a skyscraper out of fog and somehow finding that it stands.

Seen this way, Anton Zeilinger's experiments are not just exotic proofs of principle. They are reality checks for a radically non-absolutist universe, one in which assumptions, expectations, and even the observer's presence may be part of the machinery. It is the laboratory revival of an ancient intuition that reality is a dance, not a doctrine and that what you see depends on where you stand, and perhaps on who you are.

In a wry twist on Einstein's famous complaint, Stephen Hawking once remarked:

> 'Not only does God play dice, but . . . he sometimes throws them where they cannot be seen.'

Experiments have since pushed quantum behaviour well beyond elementary particles. In 2019, physicists at the University of Vienna showed that quantum interference is not limited to the tiniest scales. They demonstrated it in molecules containing more than 2,000 atoms, large enough to be visible under an electron microscope. This was not just a clever lab trick. It suggested that quantum phenomena do not suddenly disappear as systems grow larger. They become harder to observe, yes, but they do not vanish. From this view, classical physics begins to look less like a separate regime and more like a rough approximation that works for big, messy systems.

What followed was stranger still. Consider the delayed-choice quantum eraser, inspired by John Wheeler's audacious ideas from the 1980s. In these experiments, a choice made in the present about how to measure a particle seems to influence how that particle behaved in the past. This does not violate relativity, and no signal travels faster than light, but it does shatter our classical intuitions about time and causality. In the quantum world, particles do not possess definite properties until they are measured. Reality resembles a cosmic game of choose-your-own-adventure, where observation does not merely reveal what happened, but helps determine what can be said to have happened at all.

Today, quantum mechanics is no longer confined to blackboards and thought experiments. It now drives the frontiers of technological innovation. Quantum computing, with its reliance on superposition, entanglement, and interference, tackles problems so complex that classical machines simply cannot keep pace. Then there is Shor's algorithm, which can factor enormous numbers at speeds that threaten to render current encryption obsolete, something even the greatest mathematicians of the past could scarcely have imagined.

What truly distinguishes quantum thinking, however, is not just its computational power but its rejection of absolutes. It does not insist on a single, fixed reality. Instead, it reveals a world in flux, shaped by context, interaction, and observation. This shift is not merely science refining its tools, but a movement that resonates with philosophical traditions in which knowledge was understood as layered, relational, and continuously unfolding. One hears echoes of the Rigvedic idea of ṛta, or the Upaniṣadic insight that grasping the underlying unity allows the many to be seen clearly.

In this sense, we are not only building quantum systems to describe reality. We are building them to participate in it. Somewhere between the indeterminacy of a quantum bit and the clarity offered by ancient insight, a new understanding is taking shape, one that does not seek to close questions with final answers, but to hold them open, inviting deeper exploration.

The Quantum Universe

What if the universe did burst into existence with a bang, but that bang was only the aftershock of a quantum wink? Welcome to quantum cosmology, a mind-bending fusion of particle physics and the story of everything. In this view, the Big Bang is not merely a fiery starting pistol. It may have been a subtle murmur from the quantum vacuum, a form of 'nothing' that is anything but empty.

According to several modern theories, the cosmos could have emerged from a quantum fluctuation, a fleeting ripple in a field so fundamental that space and time themselves appear as late arrivals. From a sea of pure potential, governed by uncertainty and untouched by classical laws, the universe may have simply tunnelled into existence. No prior space, no prior time, not even matter was required. This is not science fiction but a serious proposal advanced by physicists such as Alexander Vilenkin.

Strikingly, these ideas echo some of humanity's oldest philosophical insights. Consider śūnyatā from Indian and Buddhist thought, a concept often translated as 'emptiness' but better understood as a dynamic and generative void. It is not absence but possibility. The Buddhist image of the universe unfolding from formless potential aligns uncannily with quantum cosmology's account of creation from the vacuum. Ancient Indian cosmologies likewise described reality as emerging cyclically from a timeless, spaceless substratum. Against this philosophical backdrop, even the strange physics of quantum foam begins to feel less alien and more like a scientific articulation of that same generative void.

Quantum foam refers to the restless agitation of spacetime itself, a seething ocean of microscopic energy fluctuations that flicker in and out of existence at the smallest scales imaginable.

This perspective shift does more than revise physics textbooks. It gently resurrects an ancient intuition long hiding in plain sight: reality is not fixed or final. It is a play, a līlā, unfolding in countless acts. When you gaze at the night sky, you may be looking not at something solid and settled, but at a ripple in the void, a cosmic improvisation born from a field that had nothing except the potential for everything.

In the early 1980s, Alan Guth stumbled onto another revolutionary idea while grappling with problems in the Big Bang model. His theory of cosmic inflation proposed that in the universe's first unimaginably brief moment, it did not merely expand. It ballooned exponentially, driven by a mysterious quantum field known as the inflaton. In a fraction of a second, a speck smaller than an atom could have swollen into a cosmos the size of a grapefruit. And as the implications of this idea unfolded, its true significance became impossible to ignore.

Inflation explained why the universe appears so smooth and flat on large scales, like a perfectly iced cake with only the faintest texture. It also explained the origin of cosmic structure itself. Tiny quantum fluctuations in the inflaton field were stretched by inflation and later became the seeds of galaxies, stars, and planets. In this sense, the fingerprints of quantum uncertainty are written across the entire night sky.

The story grows even stranger with Lee Smolin's proposal of cosmic natural selection. In this framework, black holes are not merely endpoints of stellar collapse. They may be the birthplaces of new universes. Each black hole could spawn a cosmos with slightly altered physical laws, much like genetic mutations in biology. Universes that produce more black holes would generate more descendants, gradually dominating the multiversal lineage.

If Smolin is correct, the so-called constants of nature, such as the strength of gravity or the mass of the electron, may not be eternal truths. They could be evolving tendencies, shaped by an ongoing process of cosmic selection. This is not just non-absolutism; it is cosmic improvisation taken seriously.

From Vedic myths of endless cycles of creation and dissolution to modern visions of an evolving multiverse, a pattern begins to emerge. The universe may not be a static sculpture carved once and for all. It may be more like an improvisational jazz performance—an example I have used earlier and will return to in the chapters ahead—adapting, experimenting, and riffing its way forward. In this sense, the cosmos is not merely expanding; it is trying things out.

In the late 1990s, the universe delivered another surprise. Astronomers studying distant supernovae expected to find that cosmic expansion was slowing under gravity's pull. Instead, they discovered the opposite. The expansion is accelerating, as if the universe has its foot pressed harder and harder on the accelerator. The cause of this runaway behaviour is a mysterious phenomenon called dark energy. We do not know what it is, only that it accounts for nearly 70 per cent of the universe and acts as though space itself is eager to expand. This highlights the fact that space is not a neutral stage. It has dynamics of its own.

There is an intriguing historical footnote to this story. To explain the phenomenon, physicists dusted off Einstein's old cosmological constant, a term he once dismissed as his 'biggest blunder.' History, it seems, may be granting him a quiet redemption.

On the whole, we have entered an era in which observers matter, space behaves, and unseen forces revise what once appeared to be permanent laws. From the quantum jitters that may have ignited the Big Bang to the suggestion that reality is participatory at its core, we are confronting a cosmos that is fluid, responsive, and still unfinished. As science reaches this frontier, it begins to brush against philosophy. What emerges is not a universe that simply exists, but one that is continually becoming, shaped by uncertainty, observation, and the deep structures that bind everything together.

The Universe Defies Absolutism

Let's take a quick overview of everything we have covered so far. Nearly a century of quantum physics has uncovered a disquieting truth: reality, at its core, resists rigid and absolutist definitions. This is perfectly illustrated by Heisenberg's uncertainty principle, in which trying to pin down both the position and the speed of a particle is like trying to catch a shadow. The more precisely you grasp one, the more the other slips away. Add to this the broader quantum landscape, where fields fluctuate, particles behave like waves, and certainty gives way to probability. Even more speculative frameworks, such as string theory, push this strangeness further by proposing a universe with far more dimensions than everyday experience can accommodate.

Moreover, this resistance to fixed rules does not invite despair or relativism. On the contrary, it calls for intellectual humility. Our scientific theories, however powerful, are models rather than mirrors of reality. They map the terrain, but they are not the terrain itself. David Bohm repeatedly warned against confusing our conceptual frameworks with the world they aim to describe. Trying to reduce reality to a model is like trying to capture the taste of chocolate using only the word 'sweet.' By doing so, something essential is always left out.

Another crucial point to recognise is that quantum mechanics does not claim reality to be illusory or arbitrary. Experiments remain precise, repeatable, and empirically robust. The surprise lies elsewhere. The properties of quantum systems do not simply exist in isolation, waiting to be revealed. They emerge in relation to how we observe and measure them. Observation is not a passive act of discovery; it is part of the phenomenon itself. Open the box, and what you find depends on how you open it.

This does not mean that quantum mechanics claims anything goes. Instead, it points to something subtler and more profound. Reality is real, but it is contextual. It unfolds through interaction.

Certainty gives way to probability, and permanence gives way to responsiveness. This is why quantum theory remains revolutionary. It does not just revise equations; it invites us to rethink what it means to know the world at all.

This revolution began quietly in the early twentieth century with Max Planck's discovery that energy comes in discrete quanta. From that modest insight flowed transformations that now ripple far beyond physics. Quantum ideas have reshaped computing, biology, and even how we think about the mind. It is as if the universe revealed a hidden playlist, and quantum mechanics turned out to be the DJ.

Consider quantum computing, an idea once confined to thought experiments. Classical computers operate through deterministic steps, processing bits that are either zero or one. Quantum computers exploit superposition and entanglement, allowing qubits to explore many possibilities at once. This makes certain problems, such as factoring large numbers, simulating molecules, or optimising complex systems, tractable in ways classical machines cannot match.

Biology, too, has entered the quantum conversation. Experiments suggest that photosynthetic organisms exploit quantum coherence to transfer energy with remarkable efficiency. Plants and algae appear to take advantage of quantum effects that engineers are only beginning to harness. Nature, it seems, learnt the rules long before we did. Even cognitive science has begun to explore quantum-inspired models of the mind. Rather than functioning as purely linear, deterministic machines, our brains may operate in ways that are probabilistic and context-sensitive. Thought, memory, and decision-making might follow patterns that resemble quantum processes more than classical logic.

Here, the common thread running through all these fields is that each has come up against the limits of rigid, absolutist thinking. Whether in quantum computing, cognitive science, or our evolving understanding of the mind, progress has come from accepting uncertainty rather than resisting it. Just as quantum mechanics operates through probability, context, and observation, these disciplines are moving away from fixed answers and towards adaptability and nuanced reasoning.

This shift matters far beyond the laboratory. As the world confronts climate change, rapid technological disruption, and mounting resource pressures, absolutist thinking increasingly reveals its weakness. Rigid solutions tend to collapse when confronted with new data or changing conditions. A quantum-informed mindset, by contrast, emphasises flexibility, responsiveness, and evolution. It does not surrender to chaos or drift into nihilism. Instead, it looks for creative ways to function within a world that is constantly in motion. And this is where quantum mechanics truly earns its philosophical weight. It does not promise final answers. If anything, it continually reminds us of how much remains unknown. Richard Feynman once remarked, 'I think I can safely say that nobody understands quantum mechanics.' This was not an admission of defeat. It was a recognition that deep understanding often comes hand in hand with humility. The more we learn, the stranger and richer the universe reveals itself to be.

To close this journey, it is fitting to return to Niels Bohr, whose principle of complementarity captures the heart of quantum non-absolutism. Bohr famously observed:

> 'The opposite of a correct statement is a false statement. But the opposite of a profound truth may well be another profound truth.'

In quantum physics, this insight finds concrete expression. Light behaves as both a particle and a wave, yet neither description alone exhausts its nature. Each framework reveals something essential while leaving something out. Together, they offer a deeper, though still incomplete, picture.

This logic extends into every aspect of life. Human experience itself is structured by apparent contradictions: joy and sorrow, certainty and doubt, order and chaos. These opposites do not cancel each other out. They enrich our understanding precisely because they coexist. Bohr's insight reminds us that reality is not a rigid machine governed by single, exclusive truths, but a dynamic web of possibilities shaped by context and perspective.

As we push deeper into quantum frontiers, from the search for quantum gravity to the strange implications of quantum biology, non-absolutism remains a crucial guide. The universe does not behave like clockwork. It unfolds as a network of potentialities, continually shaped by interaction and observation. And perhaps the most important lesson lies here. Just as classical physics eventually gave way to the quantum revolution, future discoveries may force us into an even more interconnected and counterintuitive view of reality. Embracing non-absolutism does not mean abandoning understanding. It means learning how to think with greater flexibility, humility, and openness.

In the end, the point is not to collect final answers, but to learn how to live with evolving ones. Reality, like quantum mechanics itself, is not a destination. It is an unfolding dance of probabilities, and learning to move with it may be the deepest form of understanding we have.

'Reality is not a quality you can test with litmus paper.'

— **Stephen Hawking**

chapter 3

the uncertain fabric of space and time

For centuries, humanity lived with a comforting assumption that time flowed like a river, always forward, always uniform, and the same for everyone, everywhere. Space, meanwhile, was the solid stage on which the drama of existence unfolded, fixed, absolute, and unaffected by anything happening within it. These ideas were so deeply woven into common sense that questioning them felt almost absurd, like doubting the blueness of the sky or the pull of gravity.

Isaac Newton gave these intuitions their most powerful and enduring form. In his formulation, the universe was predictable, deterministic, and complete. Space, he declared, was 'absolute, always similar and immovable,' while time 'flowed equably without regard to anything external.' These were not minor technical assumptions. They were the bedrock on which reality itself seemed to rest. However, by the late nineteenth century, this bedrock began to crack. Experimental results started to resist Newton's tidy picture, especially in the emerging field of electromagnetism. Physicists believed that light travelled through a hidden medium called the luminiferous aether, much as sound needs air to propagate. This invisible substance was meant to provide an absolute reference frame for motion through space.

The decisive blow came in 1887, when Albert Michelson and Edward Morley designed an experiment to detect earth's motion through this aether. Their apparatus was exquisitely sensitive. If the aether existed, earth's movement through it should have been unmistakable. Instead, the result was null. No motion. No drift. Nothing at all. It was like measuring the flow of a river and discovering the water was perfectly still. This outcome struck at the heart of Newtonian space and the aether hypothesis alike. What had seemed solid and unquestionable now looked disturbingly fragile. It became evident that the universe was not behaving as expected.

At first, the revolution this triggered was quiet, almost hesitant. But the anomalies kept accumulating. Mercury's orbit refused to line up precisely with Newton's predictions, deviating by a small but stubborn amount, like a dancer missing a beat despite flawless rehearsal. Radioactivity posed an even deeper challenge. Energy appeared to vanish and reappear in ways that violated the classical principle of conservation, as though nature itself had started to ignore its own rules. Then there were Maxwell's equations, which brilliantly unified electricity and magnetism. They worked with astonishing precision yet carried a radical implication. According to these equations, light travels at the same speed regardless of the motion of the observer. Whether you stood still or raced forward, light behaved as if your motion did not matter at all. This flatly contradicted Newton's framework, where space and time provided an absolute backdrop against which all motion was measured.

These puzzles were not minor technical glitches waiting for clever fixes but warning signs of something deeper. The classical concepts of space and time, long treated as unshakable foundations, were beginning to wobble. If space and time were characters in this unfolding drama, their identities were about to be radically rewritten.

The Annus Mirabilis

In 1905, a young patent clerk named Albert Einstein, who until then was more familiar with the paperwork of other people's inventions than with academic prestige, decided to make physics far more interesting. At just twenty-six, he published four papers that would permanently change the foundations of science. One introduced the quantum theory of light, planting the seeds of modern quantum physics. Another offered convincing evidence for the existence of atoms. A third explained the jittery motion of particles suspended in fluids, now known as Brownian motion. But it was the fourth paper, *On the Electrodynamics of Moving Bodies*, that detonated the real conceptual charge.

Einstein did not try to patch up the growing contradictions between Newton's mechanics and Maxwell's equations of electromagnetism. Instead, he questioned the assumptions beneath them. What if space and time themselves were misunderstood? What if the speed of light was not a technical nuisance, but the one unshakeable fact around which everything else had to bend? On this basis, he proposed something radical: the speed of light is the same for all observers, no matter how fast they are moving. This was not a minor adjustment. It overturned centuries of thinking about motion, space, and time. If light's speed was fixed, then space and time could no longer be absolute. They had to become flexible.

From this single idea flowed consequences that still feel strange today. Imagine an astronaut travelling past earth at an enormous speed. To observers on earth, the astronaut's spacecraft would appear slightly squashed along the direction of motion, as though space itself had compressed it. The astronaut's onboard clock would also seem to run slow, so that less time passes for her than for those watching from earth. Yet from the astronaut's perspective, nothing about her ship or their clock has changed. Instead, it is earth that appears contracted and earth's clocks that seem delayed.

But there is more: even simultaneity collapses in such cases. Suppose two lightning bolts strike the front and back of a moving train. An observer standing on the platform might judge the strikes to be simultaneous. But a passenger seated inside the speeding train, moving towards one flash and away from the other, would conclude that one strike occurred before the other. What is simultaneous for one observer is not necessarily simultaneous for another.

These were not just clever thought experiments. Einstein's equations showed precisely how time dilates and lengths contract at high speeds. Mass, too, was no longer fixed. As an object's speed increased, so did its resistance to acceleration. Motion altered matter itself.

Perhaps most striking was the introduction of a universal speed limit. Nothing could travel faster than light. At roughly 300,000 kilometres per second, it was not merely very fast; it was the maximum speed allowed by the structure of reality. Moreover, Einstein also revealed another astonishing implication: mass and energy are interchangeable. The now-famous equation, $E = mc^2$, appeared quietly, without fanfare. Yet it implied that a tiny amount of mass could release an enormous amount of energy. Decades later, this insight would power both nuclear reactors and nuclear weapons, a reminder that abstract ideas can carry very real consequences.

Einstein achieved all this without leaning on new experiments. The technology of his time could not yet test most of these effects. Instead, he relied on conceptual clarity and a willingness to rethink first principles. His special theory of relativity rested on a simple demand that the laws of physics must be the same for all observers moving at constant speeds relative to one another. With that insistence, Einstein dismantled the old notions of absolute space and absolute time. They were no longer the fixed stage on which events unfolded, but dynamic participants in the cosmic drama. In doing so, he did more than rewrite physics. He forced us to rethink what we mean by time, space, and reality itself.

Few people can claim to have given the universe an existential crisis. Einstein did it before turning thirty.

When Einstein first unveiled his special theory of relativity, the response was far from celebratory. The scientific community was less impressed than puzzled. The prevailing reaction was closer to 'Who is this guy?' than 'He might be right.' But the universe has a habit of siding with accuracy over authority, and it did not take long for Einstein's predictions to prove themselves. It later emerged that his claims were not speculative leaps, but precise and testable.

In 1908, Max von Laue became one of the first to provide experimental support by showing that fast-moving electrons possessed more mass than identical electrons at rest. This was not philosophical musing or mathematical cleverness; it was concrete evidence that nature behaved in ways classical physics could not explain.

Even time itself soon followed suit. In 1938, physicists Herbert Ives and G. R. Stilwell measured the Doppler shift in rapidly moving hydrogen atoms and confirmed a key prediction of relativity: time runs differently for objects in motion. Clocks moving at high speeds tick more slowly, a result as strange as it was undeniable.

What had once sounded like a thought experiment now stood on firm experimental ground.

By around 1915, most physicists had accepted Einstein's ideas, if not always comfortably. However, the public was far slower to come around. The notion that time could slow down simply because something was moving quickly, or that space itself could stretch and contract, felt deeply counterintuitive. People were not ready to believe that a wristwatch might lose time just for going too fast.

Yet the most disorienting idea was neither time dilation nor length contraction. It was the relativity of simultaneity, a phrase that sounds abstract but cuts straight to the heart of how we experience reality. Consider two distant fireworks launched from opposite sides of a valley. An observer standing midway between them may see the explosions at the same moment and conclude that they were simultaneous. But another observer, riding in a moving car towards one of the fireworks and away from the other, will see the nearer explosion first and the farther one slightly later.

In this case, both observers are correct. What differs is their state of motion. Simultaneity, it turns out, is not universal. Events that appear to occur at the same time for one observer may not do so for another. The idea of a single, shared 'now' quietly dissolves. This was more than a challenge to common sense; it was a direct assault on the notion of an absolute present. Einstein was not only saying that time is relative, but that the very moment we call 'now' depends on where we are and how we move. There is no global snapshot of reality ticking away in perfect unison. Instead, each observer carries their own version of the present.

It is no wonder this idea caused discomfort. It unsettled philosophers and physicists alike. But Einstein was not troubled by the confusion. He pressed on, confident that nature would vindicate him. And, as it turned out, the universe eventually did.

Minkowski's Elegant Unification

In 1908, Hermann Minkowski, who had once dismissed his former student Albert Einstein as something of a 'lazy dog' in mathematics, experienced an insight that would redefine modern physics. Stepping onto the lecture stage with theatrical confidence, he delivered a declaration that sounded less like mathematics and more like prophecy:

> 'Henceforth space by itself, and time by itself, are doomed to fade away into mere shadows, and only a kind of union of the two will preserve an independent reality.'

This was not a minor technical adjustment. It marked the birth of the spacetime revolution. Minkowski realised that Einstein's equations made far more sense when space and time were treated not as separate entities, but as parts of a single, unified structure. Every event, every moment, and every location could be placed within one four-dimensional continuum: spacetime. In this new picture, space and time were no longer independent measures competing for attention. They were inseparable, each defined only in relation to the other. The appeal was not merely mathematical elegance. This shift fundamentally altered how the universe itself was understood.

At first, Einstein was sceptical of Minkowski's four-dimensional framework. It struck him as an unnecessary mathematical flourish, elegant perhaps, but not essential. Yet Einstein, always willing to rethink his position, soon recognised that Minkowski's insight was what transformed special relativity from a clever theory into a truly revolutionary view of the cosmos.

Minkowski also introduced a concept that quickly became central to relativity: the spacetime interval. Instead of measuring distances in metres and durations in seconds, the spacetime interval combined both into a single quantity. Crucially, this quantity was invariant. No matter how fast an observer moved or where they were located, everyone would agree on its value. An event that appeared separated mainly by distance for one observer might appear separated mainly by time for another, yet the spacetime interval between them remained the same. Relativity did not dissolve order; it revealed a deeper one.

Perhaps Minkowski's most powerful contribution was the light cone. This simple geometric idea provided a visual map of causality itself. The light cone shows which events can influence one another and which cannot. Events within your future light cone can, in principle, be affected by what you do now. Events outside it remain forever beyond your causal reach. No matter how fast you travel, you cannot send a signal or exert influence beyond this boundary. In this way, the light cone made visible one of relativity's most profound rules: nothing outruns light, and cause always precedes effect. Any attempt to escape this structure is met with a firm refusal from the universe itself.

Furthermore, within this framework, attempting to travel at the speed of light leads to a strange boundary known as a null interval. In simple terms, this meant that time would stop entirely. No ageing, no ticking clocks, no passage of moments at all. It is not a triumph of motion but a dead end . . . a region where ordinary experience simply disappears.

Things become even stranger if one imagines pushing beyond that limit. Faster-than-light motion would not merely break records; it would unravel causality itself. Events could appear to run backwards in time, producing paradoxes in which cause and effect lose their meaning. The kind of time-travel fantasy that sounds entertaining quickly turns into a logical disaster. Physics, quite sensibly, refuses to allow it.

An important point to note here is that these constraints are not technical inconveniences waiting to be engineered away. Rather, they are woven into the structure of spacetime itself.

Minkowski's contribution was therefore not just mathematical polish; it completed the revolution Einstein had begun.

On the whole, Hermann Minkowski was the unsung hero of the special relativity story. His synthesis was not merely a clever reformulation of Einstein's work but a deep and intuitive picture of reality's underlying architecture that physicists still rely on today. In a sense, Minkowski turned Einstein's theory into a cosmic dance, where space and time were no longer distant guests but inseparable partners, moving together in an endless, interdependent rhythm. And it all came from the same professor who once doubted Einstein's mathematical seriousness. Few academic reversals have been quite so cosmically ironic.

When Spacetime Curves

Einstein was never one to rest on his laurels, and he was not satisfied with the triumph of special relativity alone. While the theory radically reshaped our understanding of space and time, it came with a crucial limitation. It applied only to observers moving at constant speeds. The moment acceleration or gravity entered the picture, special relativity began to falter.

There was a deeper problem as well. Special relativity insisted that nothing could travel faster than light, yet gravity seemed to defy this rule. In Newton's framework, gravitational influence acted instantaneously across any distance, as though a force could leap across the universe without delay. Einstein could not accept this contradiction. If the speed of light was truly a cosmic limit, gravity could not be exempt. Solving this puzzle became his next great ambition.

The result was the general theory of relativity, completed in 1915. But before Einstein could construct the mathematics, he needed a conceptual breakthrough. He later called it 'the happiest thought of my life,' and it arrived through a simple mental experiment involving an elevator.

Imagine standing inside a sealed elevator, far from any gravitational source. If the elevator were in free fall, you would feel weightless, floating as though gravity had vanished. Now imagine the same elevator at rest on earth. You would feel your weight pressing you to the floor. Einstein realised that, from inside the elevator, there was no way to distinguish between these two situations. The effects of gravity and acceleration were fundamentally equivalent.

This insight became the equivalence principle, and it transformed gravity from a force into a feature of spacetime itself. Gravity was no longer an invisible pull acting at a distance. Instead, massive objects curve the geometry of spacetime, and smaller objects move along those curves. Planets do not orbit because they are tugged by a mysterious force, but because they follow the natural paths carved out by warped spacetime.

A common analogy helps make this idea intuitive. Picture a bowling ball resting on a stretched trampoline. The ball creates a dip in the fabric, and smaller balls placed nearby roll towards it. In a similar way, the Sun curves spacetime, and planets move within that curvature. As physicist John Wheeler later captured it succinctly: 'Matter tells space how to curve, and curved space tells matter how to move.' Gravity, then, is not a spooky influence acting instantaneously across the cosmos; it is the geometry of the universe in motion.

With general relativity, Einstein did more than refine existing ideas but redefined the relationship between matter, energy, space, and time themselves. The theory explains phenomena such as the bending of light around black holes, the slow drift of planetary orbits, and the precise timing of GPS satellites. It also resolved long-standing cosmic puzzles, including Mercury's slightly errant orbit. For decades, astronomers had observed that Mercury drifted from the path Newton's laws predicted. Einstein's theory explained the discrepancy by showing that the sun's immense mass warps spacetime, subtly reshaping the planet's trajectory. And this was not abstract theorising. It was a precise, testable result, and it immediately elevated general relativity from speculation to scientific breakthrough.

The true public turning point came in 1919, when Arthur Eddington led an expedition to test another of Einstein's predictions that light bends when it passes near a massive object. According to general relativity, starlight grazing the sun should curve as it travels through distorted spacetime. During a total solar eclipse, Eddington's team measured the apparent shift in star positions and found exactly the deflection Einstein had predicted. The result made headlines worldwide. Overnight, the former patent clerk became an international celebrity, and relativity entered the public imagination.

Less dramatic, but just as profound, was Einstein's prediction of gravitational time dilation. Time, he argued, runs more slowly in stronger gravitational fields. This idea sounds like science fiction, yet it has become a practical necessity. GPS satellites must correct for both special and general relativistic effects. Without these adjustments, navigation errors would accumulate at roughly ten kilometres per day. Modern positioning systems work precisely because spacetime behaves exactly as Einstein described.

At a deeper level, general relativity overturned the classical picture of space and time themselves. In Newton's universe, space and time formed a rigid, passive backdrop against which events unfolded. In Einstein's universe, spacetime is dynamic and responsive. It bends, twists, and even ripples in response to mass and energy. In other words, spacetime is not merely a stage; it is an active participant. Matter and energy shape the geometry of spacetime, and that geometry governs how matter and energy move. Einstein's field equations formalised this relationship, revealing a universe in which structure and motion are inseparably linked. Black holes, gravitational waves, and even the theoretical possibility of time travel all emerge naturally from this framework.

General relativity also foretold the existence of gravitational waves, ripples in spacetime produced by violent cosmic events such as merging black holes. Einstein derived them in 1916 but assumed they would be impossible to detect. A century later, in 2015, the LIGO collaboration observed these waves directly, measuring distortions in spacetime smaller than the width of a proton. It was an extraordinary technical achievement, confirming yet another bold prediction of the theory.

Ultimately, Einstein did not simply explain a few unusual phenomena. He anticipated entire chapters of the universe's story, many of which humanity is only now beginning to read. So, the next time you feel gravity anchoring you to the ground, remember that you are not being pulled by an invisible force. You are moving through the curves of spacetime, participating in a vast and elegant cosmic waltz.

A Drama of Expansion

When Einstein first applied his general theory of relativity to the universe as a whole, he ran into a profound problem. His equations refused to describe a static cosmos. Instead, they implied that the universe must either be expanding or contracting. This conclusion clashed sharply with the dominant belief of the time, shared by most leading astronomers, that the universe was eternal, unchanging, and fixed in size.

To reconcile his theory with this assumption, Einstein introduced an additional term into his equations: the cosmological constant. This term acted as a kind of large-scale repulsive effect, counterbalancing gravity and allowing the universe to remain static. At the time, it seemed like an elegant solution, a mathematical adjustment that preserved both his theory and the prevailing worldview.

Then came 1929. Edwin Hubble, through meticulous observations of distant galaxies, made a discovery that overturned the entire picture. He found that galaxies were moving away from us, and that the farther away a galaxy was, the faster it appeared to recede. Put simply, the universe was not static at all. It was expanding.

Faced with this evidence, Einstein reportedly expressed deep regret over his earlier modification. He later referred to the cosmological constant as his 'greatest blunder,' not because it was mathematically flawed, but because it reflected a failure to trust the natural implications of his own equations. Had he done so, he might have anticipated the expanding universe years before it was observed.

The realisation that the universe is expanding transformed cosmology. If space is expanding today, then in the distant past it must have been smaller, denser, and hotter. Tracing this expansion backwards leads to an extraordinary conclusion: the universe began in an extremely compact, high-energy state. This idea became the foundation of what we now call the Big Bang theory.

At first, the notion was met with resistance. Many found it unsettling to imagine that space, time, and matter all emerged from a single beginning. But as observational evidence accumulated, the Big Bang became not merely plausible, but unavoidable.

One of the most decisive confirmations arrived by chance in 1964. While investigating persistent noise in a radio antenna, Arno Penzias and Robert Wilson detected a faint, uniform background of microwave radiation coming from all directions in the sky. This cosmic microwave background turned out to be the lingering thermal afterglow of the early universe, a relic of the Big Bang itself. The signal had been predicted by theorists such as George Gamow, yet its accidental discovery surprised everyone. Penzias and Wilson were awarded the Nobel Prize, and the Big Bang model was firmly established.

However, the story did not end there. The expansion revealed something even more astonishing. It was not simply galaxies moving through space. Space itself was stretching. Distances between galaxies increased because the fabric of the universe was expanding beneath them. This idea marked a radical break from classical physics. In Newton's framework, space was an inert stage on which events played out. In relativistic cosmology, space becomes dynamic. It can expand, curve, and even change the rate of its expansion over time. And just when it seemed that the universe had revealed its deepest secrets, it hinted that it was capable of far stranger behaviour still.

In the late 1990s, astronomers made a startling discovery. Despite the pull of gravity, which was expected to slow it, the universe's expansion showed no sign of deceleration. On the contrary, it was accelerating. This finding ran directly against classical intuition, which held that gravity should gradually pull matter back together. Instead, observations revealed that some unknown influence was overpowering gravity on the largest scales. This mysterious driver of cosmic acceleration came to be known as dark energy.

The discovery sent shockwaves through cosmology. Scientists were suddenly confronted with a universe dominated by an entity they could not see, measure directly, or fully explain. Ironically, Einstein's equations had already allowed room for such a possibility through the cosmological constant, the very term he once dismissed as a mistake. Moreover, general relativity also permitted different large-scale geometries of the universe. This meant that on cosmic scales, space can be flat like a plane, positively curved like the surface of a sphere, or negatively curved like a saddle. Precise measurements of the cosmic microwave background suggested that the universe is remarkably close to flat. This raised deep questions about the universe's earliest moments and why it emerged in such a finely balanced state.

To address these puzzles, physicists developed the theory of cosmic inflation. According to this idea, the universe underwent an extremely rapid burst of expansion in the tiniest fraction of a second after the Big Bang. Inflation helps explain why the universe appears flat, uniform, and structured in the way we observe today. It was an attempt to understand how the cosmos could change shape so dramatically, so quickly, and yet remain so orderly.

Although the Big Bang and inflation theories helped explain numerous cosmic phenomena, they also raised many new questions. Will the universe expand forever, thinning out into a cold, dark expanse? Will gravity eventually halt the expansion and reverse it, leading to a catastrophic collapse known as the Big Crunch? Or could the acceleration grow so extreme that space itself tears apart in a Big Rip, pulling galaxies, stars, atoms, and even spacetime itself into fragments?

We do not yet know which path the universe will take. What is clear is that we are not mere spectators. We are embedded within this unfolding drama, subject to the same cosmic forces that shape galaxies and fate itself. The universe has no final script, and its next act remains unwritten. And that uncertainty is not a failure of knowledge. It is an invitation to keep asking deeper questions about the nature of reality.

A Block Universe

At the beginning of the twentieth century, a series of theoretical breakthroughs began to undermine long-held intuitions about time itself. This included Einstein's theory of relativity, whose implication was the collapse of simultaneity. What once seemed obvious—that two events could occur at the same time for everyone—turns out not to be universal at all.

In relativity, simultaneity depends on the observer and on how fast that observer is moving. In other words, time is not shared equally by all. This idea strikes at the heart of classical physics, where space and time were assumed to be absolute and independent of the observer.

From this insight emerges a radical picture of reality known as the block universe. In this view, time is not a river flowing from past to future. Instead, the universe is a four-dimensional structure in which past, present, and future all exist together. Every moment is equally real. Nothing is coming into being, and nothing is slipping away. This idea can be illustrated with the example of a loaf of bread. Each slice represents a moment in time, yet the loaf itself exists all at once. No slice is more real than any other. Therefore, the distinction between past, present, and future arises not from the universe itself, but from how we experience it.

This idea takes us deep into metaphysics. The philosopher and mathematician Hermann Weyl captured its essence with striking clarity:

> 'The objective world simply is, it does not happen. Only to the gaze of my consciousness, crawling along the lifeline of my body, does a section of this world come to life as a fleeting image in space which continuously changes in time.'

According to this view, what we call the 'present' is not a moving boundary between what has been and what will be. It is a feature of consciousness. We experience one moment at a time, even though all moments are already there.

This is why the block universe challenges our most basic intuitions. The passage of time, the steady ticking we all feel, may not be a feature of reality at all. It may be a psychological effect.

One way to picture this is to imagine the universe as a vast Netflix series. Every episode, every moment of your life, has already been filmed, edited, and archived in flawless cosmic resolution. But you experience the story one scene at a time, believing you are writing it as you go. Your first heartbreak, your greatest triumph, even that strange dream where you are late for an exam in a room full of ducks all coexist, eternally preserved in spacetime's grand library.

Einstein, too, fully embraced this view later in his life. For him, time was no longer a flowing entity but a coordinate, woven inseparably into the fabric of spacetime. This perspective finds a quiet, moving expression in a letter he wrote in 1952 to the family of his close friend Michele Besso, shortly after Besso's death, where Einstein reflected on the illusory nature of temporal separation.

> 'Now he has departed from this strange world a little ahead of me. That means nothing. People like us, who believe in physics, know that the distinction between past, present, and future is only a stubbornly persistent illusion.'

This was not a casual musing. It marked Einstein's quiet acceptance that time, at least as we experience it, may be a stubborn illusion.

This idea is also echoed by Julian Barbour, who proposes a view in which time may not exist at all. He argues that what we call time is nothing more than a way of describing change. The ticking clock, the sense of before and after, emerges from the relationships between different states of the universe. Time, in this picture, is not an independent entity or a fundamental force. It is a byproduct of how things are arranged and how they differ from one another.

From Barbour's perspective, the universe does not evolve through time. Instead, it consists of a vast collection of 'nows,' each a complete configuration of reality. Our experience of motion and duration arises because some of these configurations contain records or memories of others. This perspective forces us to reconsider one of our deepest intuitions. The sense that the present is slipping away and that the future is approaching may be a cognitive tool rather than a cosmic truth. It helps us navigate the world, but it may not describe how reality is fundamentally structured.

However, not everyone is ready to abandon the idea of time as a fundamental aspect of reality. Physicist Lee Smolin has argued forcefully that time is real, dynamic, and essential. In his view, time is not an illusion layered on top of a static universe. It is the engine that allows novelty, evolution, and genuine change to occur. Without the passage of time, Smolin argues, the universe would be frozen, incapable of producing surprises or unfolding possibilities. In this conception, time is not a passive backdrop but an active participant in the cosmos. It shapes what can happen and makes becoming possible.

What is striking is that all these major theories dismantle the familiar image of time as a steady, external clock ticking away in the background of our lives. The implication is profound: time—the framework we have relied on for centuries to understand change, causality, and meaning—may not be the foundation of reality after all. It could be a powerful psychological and relational construct, one that helps us navigate the world but does not exist in the way we instinctively believe it does. And yet, despite all this, time remains central to human life. Our bodies age. Our biological rhythms pulse. Our societies run on schedules, deadlines, and histories.

Even if time is not fundamental to the universe, it remains fundamental to us in more ways than one. That is what makes the question so compelling. What is time, really? A deep feature of reality, an emergent property of matter, or a story consciousness tells itself to impose order on a strange and indifferent cosmos. And if time does not truly flow, if the 'now' we cling to is simply a construction of the mind, then what happens to our sense of self? Are you the person brushing your teeth this morning or the version of you winning a Nobel Prize next year? Or do both already exist, side by side, in a timeless cosmic fabric that makes our daily plans feel almost trivial?

These questions echo ideas far older than modern physics. In Hindu cosmology, *kāla*, or time, is often understood as cyclical and layered rather than linear. The *ātman*, the witnessing self, observes change without being fully bound by it. Literature, too, has long grappled with this intuition. Marcel Proust devoted an entire epic to the way memory collapses time, showing how a single taste or scent can carry the mind across decades. What he captured through art may be what physicists now attempt to describe mathematically: time not as a conveyor belt, but as a landscape the mind can traverse.

Popular culture has explored this idea as well. In the 2016 film *Arrival*, based on Ted Chiang's *Story of Your Life*, an alien species perceives all of time at once. Their language is written in circular forms that encode complete meanings in a single gesture. Learning this language reshapes the human protagonist's perception of time itself. It is a cinematic nod to the block universe, and a reminder that our experience of time may be deeply tied to cognition and language rather than to the structure of reality alone.

All of this forces us to rethink the canvas of our existence. The boundaries we take for granted, past, present, and future, may be little more than perspective effects. It is like mistaking a Möbius strip for having two sides when it is, in fact, a single continuous surface.

Seen this way, time becomes more than a cosmic metronome. It becomes a mirror, reflecting not just when we live, but how we engage with reality. Whether time flows like a river, lies frozen like a glacier, or dissolves into an illusion altogether, our relationship with it shapes everything we care about: love, loss, ambition, and identity.

So perhaps the deepest question is not whether time exists. It is whether we are willing to imagine ourselves without it. Because once we loosen our grip on ticking clocks and countdowns, we may glimpse something more enduring, something that does not merely rearrange our calendars but reshapes what it means to exist at all.

Relative Yet Objective

A common misconception, popular at dinner parties and on edgy T-shirts, is that Einstein's theory of relativity means everything is relative. Cue someone sipping wine and declaring, 'Well, truth is just a matter of perspective, isn't it? It's all relative.' But the irony is that if Albert Einstein were at the table, he would probably adjust his bow tie, sigh, and say, 'Nein. You have got it wrong.'

The confusion lies not in what Einstein said, but in how the word 'relativity' is usually heard. When people hear the term, they often imagine a universe gone soft, one where up might be down, clocks lie, and nothing stays still long enough to matter. Max Born, one of Einstein's more mathematically minded contemporaries, thought the name itself encouraged this misunderstanding. If he had his way, we would be calling it the theory of invariance. It lacks the flair of relativity, but it gets closer to the heart of the idea. Born's point was subtle and revolutionary. Relativity is not about everything being relative; it is about identifying what is not. The equations that govern motion, energy, and causality hold firm and this is the quiet elegance of relational objectivity.

Thus, it is important to understand that Einstein did not dismantle reality but refined it. Philosopher Hilary Putnam later expressed a similar sentiment when he remarked, 'Einstein didn't kill objectivity; he reimagined it.' In other words, he did not turn the universe into a cosmic free-for-all where every opinion counts as its own version of truth. What he showed instead was something subtler and far more powerful. That is, your measurements of space and time can differ depending on how fast you are moving or how strong gravity is where you stand, but the laws that govern those measurements do not budge.

In this view, Einstein's universe is less an anything-goes and more like a finely tuned orchestra. Each observer plays their own melody of space and time, but everyone follows the same rhythm section: the laws of physics. This insight is not just theoretical elegance; it has practical consequences. Take GPS, for example. That quiet constellation of satellites that tells your phone where the nearest pizza joint is sits about 20,000 kilometres above earth. Up there, time behaves differently. General relativity tells us that clocks tick faster in weaker gravitational fields, and gravity is slightly weaker at satellite altitude than it is on earth's surface. As a result, those orbiting clocks run a bit faster than the ones in your pocket.

But that is only half the story because special relativity also comes into play in this scenario. Since GPS satellites are moving at over 14,000 kilometres per hour, their clocks run slower from our perspective due to their speed. Combine these two effects and things get tricky. If engineers failed to correct for both, your location could drift by nearly ten kilometres every single day.

The solution here is pure Einsteinian. Engineers explicitly build relativistic corrections into the system, carefully synchronising satellite clocks with those on earth. Every time your phone locks onto your location, it isn't just reading signals; it is reading relativity. Einstein, it turns out, helps you find tacos at 2 a.m.

Relativity also shows up in far more dramatic settings. In the European countryside, beneath the border between France and Switzerland, the ground hums not with earthquakes but with human curiosity. There you will find the Large Hadron Collider at CERN, a 27-kilometre underground ring where scientists accelerate subatomic particles to speeds unimaginably close to light. At these velocities, special relativity takes over completely. As particles accelerate, they behave as if their mass increases. A proton at rest is unremarkable. But push it near light speed and it resists acceleration like a stubborn bowling ball.

To control these particles, physicists rely on magnetic fields thousands of times stronger than earth's, generated by superconducting magnets cooled to near absolute zero. Steering them is less like driving a car and more like guiding a raging bull with dental floss. The precision has to be so extreme that engineers need to account for the moon's gravitational pull as well. Yes, the same moon that disrupts your beach plans also nudges particle beams. And when CERN announced the discovery of the Higgs boson in 2012, briefly turning physicists into reluctant celebrities, relativity was quietly holding the entire operation together.

Perhaps the most poetic part is this: Einstein never lived to see how his ideas became the scaffolding of modern technology. No smartphones, no GPS, no viral dance trends explaining spacetime. And yet, every time your location pings accurately and every time particles collide at staggering energies, his equations are at work.

Moreover, particle accelerators require enormous amounts of energy to propel particles to such speeds, which means their design depends directly on understanding the relationship between mass and energy. Here again, Einstein comes to the rescue. His equation, $E = mc^2$, shows that mass and energy are interchangeable—an insight that makes modern high-energy physics possible.

Particle accelerators and GPS satellites may sound futuristic, but relativity also shows up in places far more familiar. Take gold, for instance. The warm yellow glow of gold jewellery is not only a matter of chemistry or ancient alchemy. It is partly a consequence of relativity. Gold is a heavy element, and in heavy atoms, the inner electrons move so fast that relativistic effects become significant. These effects alter how gold absorbs and reflects light, shifting its colour towards that distinctive golden hue.

Gold is not alone in this. Other heavy elements such as platinum, mercury, and uranium also exhibit relativistic effects that change their physical and chemical properties. These effects matter well beyond aesthetics. They also influence how catalytic converters reduce harmful emissions in cars and affect the behaviour of materials used in nuclear power generation. Shiny jewellery and clean energy, it turns out, share the same scientific foundation: relativity!

Zoom in even further, to the microscopic world of semiconductors and there, too, you will find relativity at play. These tiny components power everything from smartphones to refrigerators. Even modern microchips use similar principles. The interesting fact here is that as devices shrink and performance increases, the electrons inside them move faster and occupy energy levels that require both quantum mechanics and relativistic corrections to describe accurately. Therefore, engineers designing today's chips must account for how electrons behave when their speeds and energies push into regimes where relativity matters.

The same relativistic limits that govern electrons in silicon also scale up to shape the global technologies we rely on every day. Nowhere is this more evident than in the digital world held together by the internet—an invisible web spanning the planet. At its core are fibre-optic cables, carrying information as pulses of light that race across continents and oceans. These signals move astonishingly fast, but not infinitely so. Relativity places a firm ceiling on their speed: light travels at about 300,000 kilometres per second, and that speed is the same for everyone, regardless of motion or perspective. This simple constraint has far-reaching consequences for communication, navigation, and computing. Even in an age of instant messages and streaming data, the universe still insists on playing by its own rules.

Those rules become especially apparent in something engineers call data synchronisation. When you send an email, stream a video, or order pizza online, information does not flow as a single stream but as packets that must arrive in the correct sequence, like steps in a carefully choreographed dance. Because even light takes time to travel, tiny delays inevitably creep in. Engineers must therefore design networks that anticipate and correct for these lags, measured in nanoseconds. They may seem negligible, but they are a direct consequence of the cosmic speed limit revealed by Einstein. Much like your morning espresso quietly keeps you alert, relativity works in the background, silently ensuring that your data arrives in order.

Thus, relativity's importance cannot be overstated, for it extends far beyond black holes and distant galaxies and quietly shapes everyday life, from the accuracy of your watch to the materials in jewellery and the design of modern gadgets. Richard Feynman captured this sentiment by pointing out how Einstein's insights are embedded in the fabric of the world in ways both subtle and profound when he famously said:

> 'For over 200 years, the equations of motion enunciated by Newton were believed to describe nature correctly, and the first time that an error in these laws was discovered, the way to correct it was also discovered. Both the error and its correction were discovered by Einstein in 1905. Newton's Second Law, which we have expressed by the equation $F = \frac{d(mv)}{dt}$ was stated with the tacit assumption that is a constant, but we now know that this is not true, and that the mass of a body increases with velocity. In Einstein's corrected formula has the value $m = \frac{m_0}{\sqrt{1-v^2/c^2}}$ where the "rest mass" represents the mass of a body that is not moving and is the speed of light, which is about 3×10^5 kmsec^{-1} or about 186,000 mi·sec^{-1}.'

Another fact that is not widely recognised is that Einstein's famous equation, $E = mc^2$, emerged from special relativity. When we encounter this equation, our minds often jump to atomic bombs or collapsing stars, yet it is not reserved for cosmic catastrophes or Hollywood spectacles. It is woven into the fabric of everyday technology. The equation revealed that mass and energy are, at a fundamental level, the same substance expressed in different forms. Deceptively simple—elegant and compact in its symmetry—it transformed Einstein into a modern rishi of physics, who showed that reality can bend, fold, and interweave in ways that rival poetry. And the consequences of this insight permeate everyday life, from nuclear reactors lighting cities from Tokyo to Toronto to Tirunelveli, to medical imaging, satellite navigation, and the quiet precision of the devices we rely on without a second thought.

Perhaps the deepest legacy of relativity lies not in devices or power plants, but in the questions it leaves behind. How should we use knowledge that can alter the fate of civilisations? What responsibility follows once we glimpse how the universe truly works? In a world obsessed with speed, relativity reminds us that speed is not absolute. What appears fixed may be moving, and what seems solid may be a ripple in a vast field of energy. It whispers a strange cosmic wisdom: light can bend, time can stretch, and the greatest truths may lie not only in what we measure, but in how we choose to act on what we know.

This idea was genuinely shocking to the early twentieth century, a world that had only just begun to loosen its grip on rigid hierarchies and absolute certainties. Before Einstein, Newton's space and time were treated almost like characters from classical mythology: eternal, immovable, and unquestioned. Einstein overturned this picture. Objectivity, he argued, does not require a universal stopwatch or a sacred ruler. Instead, it requires invariants, deep structural relationships that remain unchanged even as perspectives shift. So, in place of isolated absolutes floating in a vacuum, we got a web of relations. Space and time turned out to be dynamic participants rather than silent backdrops. They behaved less like stiff Enlightenment abstractions and more like living characters whose roles depend on context. What matters, then, is not the disappearance of certainty, but its transformation.

In conclusion, clocks can indeed disagree, rulers can mislead, and simultaneity can slip through your fingers. Yet beneath this shifting surface, the universe continues to speak in a language of consistency and constraint. The same principles that have governed space and time since the Big Bang still hold, binding observers together within a shared framework, even when their experiences diverge.

The End of the Universal Present

With the many radical developments at the beginning of the twenty-first century, physicists found themselves in an unusual position. The old arguments about whether space and time are absolute had already been settled in practice, yet something deeper remained unresolved. Space and time still appeared in our equations, still structured our experience, and still seemed unavoidable. But as researchers pushed further into the foundations of physics after 2000, a striking pattern emerged: every serious attempt to understand reality at its deepest level made space and time look less solid, less universal, and less fundamental than anyone had expected. What began as a refinement of Einstein's ideas slowly turned into something more radical. The question was no longer whether simultaneity depends on the observer. The question became whether simultaneity exists at all, except as a convenient approximation.

One of the clearest signs of this shift came from how physicists began to think about time in theories of gravity. In everyday life, time feels global and shared. We assume that the universe can be divided into a sequence of moments, like frames in a film, even if different observers disagree slightly about their timing. But after the turn of the millennium, work in gravitational physics increasingly emphasised that there is no natural way to make such a division. To describe change in the universe, physicists must choose a way of labelling events as 'earlier' or 'later,' but the theory itself does not tell them which choice is correct. Many different choices work equally well, and none lead to different physical predictions. Over time, this freedom stopped looking like a technical nuisance and started looking like a message: if the laws of physics do not care how you define a global time, then global time may not be a real feature of the universe. It may be more like a coordinate system on a map—useful but not part of the territory.

In general relativity, physical predictions must be independent of the choice of coordinates (diffeomorphism invariance). While we can compute observables in specific situations (for instance, planetary distances), there is no known complete set of realistic observables that fully captures the theory's gauge-invariant content. Previous attempts either require unrealistic matter fields (e.g., dust, scalar fields) or are mathematically contrived and physically impractical. For example, in the landmark paper 'GPS observables in general relativity' by Carlo Rovelli, he shows that four satellites broadcasting proper time signals define a physical coordinate system in which the metric becomes measurable and evolves locally, offering a concrete and realistic solution to the problem of observables in general relativity.

This realisation has profound consequences. If there is no preferred way to divide the universe into moments, then there is no single 'now' shared by everything that exists. The present becomes something local, tied to particular systems and observers, rather than something that stretches across the cosmos. From this perspective, asking what is happening everywhere at this very moment is like asking which direction is truly north in empty space. The question assumes a structure that nature does not provide.

An even more dramatic challenge to absolutism emerged when physicists tried to combine gravity with quantum theory. Many approaches developed after 2000 suggest that spacetime itself is not a basic ingredient of reality. Instead, the most fundamental level may consist of networks of relationships. Space and time, in this picture, are not the stage on which physics happens but a rough sketch that emerges when these relationships are viewed at large scales. To name a few such approaches, we have the causal set theory, loop quantum gravity, and quantum information-based approaches. In causal set theory, the universe is described as a network of events ordered only by causal influence, with no underlying spacetime. In loop quantum gravity, geometry itself is built from discrete relational networks rather than existing in advance. In quantum-information approaches, spacetime emerges from patterns of entanglement linking quantum systems.

What is striking about these ideas is how completely they undermine the notion of simultaneity. At the fundamental level, the theories often say nothing about events being 'at the same time.' Some events can influence others; some cannot. Beyond that, the theory remains silent. If two events are not connected in any way, there is no fact of the matter about their temporal relationship. They are neither simultaneous nor ordered in an absolute sense. The familiar idea of a shared present appears only later, when many events are grouped together and described approximately, much as temperature emerges from the motion of countless particles, even though no single particle has a temperature of its own.

Around the same time, physicists also began rethinking the role of the observer. As we have discussed, in older theories, observers were treated as idealised, almost abstract entities—perfect clocks and rulers floating outside the system being studied. But quantum theory does not allow such a clean separation. Observers are physical systems too, subject to uncertainty and quantum effects. After 2000, researchers started taking this even more seriously and asked what happens when the observer's frame of reference is itself part of the quantum world. Prof. Vlatko Vedral has recently given a rather radical premise. In his view, the universe is made of quantum numbers—also called q-numbers—which are mathematical objects that encode probabilities and possibilities, like the chances of different outcomes in a quantum experiment.

Unlike classical numbers, which tell us exact things (like this ball is 7 meters away), q-numbers describe superpositions, where many possibilities coexist until they interact or become linked. In this picture, what we call particles are not fundamental and instead emerge from the interactions of quantum fields. Even ghost particles that we cannot detect directly are considered real if they can influence or become entangled with other systems.

Measurement, Vedral says, is just one system becoming entangled with another. There is no need for a conscious observer to collapse a wavefunction; reality is fully described by these interactions. Space and time, in this view, are not basic ingredients of the universe; they are bookkeeping tools that help us describe how q-numbers relate. In other words, the real fabric of reality is the web of quantum relationships itself, from which all the familiar features of our world emerge.

Such a premise turns out to be deeply unsettling for any remaining form of absolutism. In some approaches like quantum reference frames, indefinite causal order, and relational quantum mechanics, even the order of events can depend on the observer in a way that goes beyond Einstein's relativity. Two events may occur in one order according to one observer, and in a different order according to another, without there being a single, underlying sequence that both descriptions are approximating. This is not because we lack information but because the theory itself does not single out a unique order. Time, in these cases, behaves less like a universal river flowing from past to future and more like a relationship that only makes sense once you specify who or what is doing the observing.

Time, in many of these views, emerges only when systems are looked at in certain ways, often involving large numbers of components and irreversible processes. The familiar flow from past to future arises not because the universe is built on time but because certain physical situations make it natural to describe change as a sequence. This idea can be hard to accept because our experience of time feels so immediate and unavoidable. But physics has faced similar surprises before. Solidity feels fundamental until we learn about atoms. Heat feels basic until we understand molecular motion. In much the same way, the flow of time may feel fundamental while actually being a large-scale effect that disappears when we look closely enough.

What is remarkable about all these developments is their consistency. Coming from different directions—gravity, quantum theory, information theory—they converge on the same conclusion. The deeper physicists probe, the less room there is for absolute space and absolute time. This is not because experiments have failed to find them but because our best theories do not need them and, in many cases, cannot accommodate them at all. And the lesson of post-2000 physics is therefore not simply that space and time are relative. It is that non-absolutism runs all the way down. The universe does not appear to come equipped with a universal present, a single timeline, or a fixed spatial backdrop.

It is important to recognise that this does not mean time is an illusion or that our everyday experience is mistaken. Local clocks work, causal order is reliable, and the distinction between before and after remains essential to how we live and do science. But these features belong to specific physical situations, not to the universe as a whole. They are real in the same way that waves are real in the ocean—stable, useful, and meaningful, but not fundamental constituents of water. In letting go of absolute space and time, twenty-first-century physics has not stripped reality of structure. On the contrary, it has revealed a richer and more flexible structure than the old absolutes ever allowed. Space and time, once thought to be the rigid framework of existence, now appear as dynamic, emergent features of a universe that is deeper, stranger, and more relational than we ever imagined.

Loop Quantum Gravity

For most of us, space feels like the ultimate background. It is the apparent nothingness in which everything unfolds, a passive stage where stars burn, planets spin, and galaxies perform their cosmic waltz. But what if space is not empty at all? What if it is not even continuous?

Loop quantum gravity offers a radical answer. In this view, spacetime is not a smooth sheet but a kind of cosmic lattice, woven from loops so small they make atoms seem enormous by comparison. To grasp just how strange this proposal is, we have to shrink our imagination to the smallest scales nature allows. At this extreme frontier—the Planck scale, roughly 10^{-35} metres—space may not flow like honey but click together like Lego. Thus, where general relativity pictures space as a silk fabric curved by mass, loop quantum gravity insists it is more like a tightly knitted textile, with each thread formed from quantised geometry.

To picture this, a helpful image is that of a bead loom rather than a painter's canvas. Each bead represents a spin network, a graph-like structure in which geometry itself comes in discrete units. As these networks evolve, they give rise to what physicists poetically call 'spin foams.'

This granularity leads to one of loop quantum gravity's most striking claims. It avoids the singularities that haunt Einstein's equations. Instead of an infinitely dense point at the heart of a black hole or at the beginning of the universe, the theory predicts a bounce. The universe contracts, reaches a minimum size, and then rebounds, like a slingshot reversing direction. In other words, the laws of physics do not break down at the extremities; they simply change regime.

But loop quantum gravity is not the only attempt to rethink the very scaffolding of reality. In a quieter corner of theoretical physics, another approach takes this reimagining even further. Causal set theory proposes an even more minimalist vision. Here, the universe is not built from space, or even from loops, but from events. Fundamental happenings are stitched together not by distance, but by cause and effect. Imagine a vast ancestral tree in which each node marks something that occurred, and each branch records what followed because of it. Spacetime then becomes a causal tapestry, closer to a Go board than a continuous scroll. This perspective has an uncanny philosophical resonance. It echoes the Buddhist *abhidhamma* view of reality as a stream of momentary dharmas, discrete yet causally linked. In such a framework, space and time emerge as convenient summaries, much like motion appears smooth in a video composed of individual frames.

If that does not stretch the imagination enough, consider asymptotic safety. Gravity, usually the troublemaker at quantum gatherings, may behave itself after all, but only at extremely high energies. This approach suggests that gravity does not spiral into infinities at small scales. Instead, it flows towards a stable fixed point, a kind of quantum calm. Like a wild river that settles into a broad delta, gravity's excesses are tamed by the structure of the theory itself.

Asymptotic safety preserves the core of general relativity while extending it into the quantum domain. There is no dramatic overthrow, only a careful negotiation at the Planck-scale frontier. Einstein, who spent his later years searching for unity beneath physics, might have found this approach quietly appealing.

Then there is Erik Verlinde, the Dutch physicist who proposed a deeply provocative idea in 2010. What if gravity is not a fundamental force at all? What if it is an emergent effect, arising from deeper principles of information and entropy? In this view, gravity resembles the smell of coffee, not a substance in itself, but a consequence of underlying processes.

Verlinde suggests that gravity emerges from the way information is arranged in space, acting as an entropic pull that nudges the universe towards more probable configurations. Thus, order arises not by decree but by statistical tendency. This idea overlaps with ancient Indian notions of ṛta, a cosmic order that is not imposed from above but emerges from balance and rhythm.

In his seminal paper, *On the Origin of Gravity and the Laws of Newton*, Verlinde writes,

> 'Newton's law of gravitation naturally arises in a theory in which space emerges through a holographic scenario. Gravity is identified with an entropic force caused by changes in the information associated with the positions of material bodies. A relativistic generalisation of the presented arguments directly leads to the Einstein equations. When space is emergent, even Newton's law of inertia needs to be explained. The equivalence principle suggests that it is actually the law of inertia whose origin is entropic.'

If Verlinde is right, then Newton's apple did not fall because of a force pulling it downward but because the universe was rearranging its information. Trippy? Absolutely. But also deeply suggestive. It points towards a deeper possibility that gravity, geometry, and quantum information may be different expressions of the same underlying structure of reality.

This brings us to one of the most intriguing ideas in modern physics: ER = EPR. At first glance, it looks like a cryptic equation, but it gestures towards a startling possibility. On one side are

Einstein–Rosen bridges, better known as wormholes. These are hypothetical tunnels connecting distant regions of spacetime. On the other is quantum entanglement, the eerie linkage between particles that remain correlated even when separated by vast distances. ER = EPR proposes that these are not separate at all, but two expressions of the same underlying reality.

This idea gained concrete footing through work by Juan Maldacena and Leonard Susskind. They showed that when charged black holes form in a magnetic field through pair production, the resulting black holes emerge entangled and crucially, a wormhole connects them. More provocatively, they suggested that this may be a universal feature of entanglement. Any entangled pair of particles, from electrons to photons, could be linked by microscopic, non-traversable wormholes at the Planck scale, the smallest meaningful unit of space. One way to picture this is to imagine entangled particles as perfectly synchronised dancers, their coordination not maintained by signals racing through space, but by hidden geometric connections beneath the surface of reality. What appears as instantaneous correlation may be geometry in disguise.

In this picture, space is no longer a passive stage on which physics unfolds. It becomes a dynamic fabric woven from entanglement itself. ER = EPR hints that geometry grows out of quantum connections, as though reality were stitched together by an invisible web. Ancient myths spoke of hidden threads binding fate and form through the weaving of the Norse Norns and the strands spun by the Greek Moirai. Modern physics, it seems, is rediscovering those motifs, translated into the language of quantum information and wormholes.

Then, in 2012, the story took a dramatic turn with the firewall paradox. Joseph Polchinski and his colleagues showed that preserving both quantum mechanics and relativity seemed to require a startling conclusion. Instead of a smooth and uneventful crossing, anything falling through a black hole's event horizon would be destroyed by a searing firewall. This conclusion shattered Einstein's vision of the horizon as a perfectly ordinary region of space. But if entangled black holes are connected by wormholes, information can pass through these hidden channels without violating the speed of light. The feared firewall dissolves into something far subtler, less a wall than a hidden passage.

From all this, we can see how black holes—long imagined as cosmic prisons—have always stood at the heart of this debate. Stephen Hawking famously stirred controversy by declaring, 'There are no black holes.' What he meant was not that black holes do not exist, but that the traditional notion of an event horizon as an absolute point of no return may be flawed. Instead, Hawking proposed the idea of an apparent horizon, a temporary boundary where matter and information are trapped for long periods but not forever. This move was aimed at resolving the information paradox, the clash between Einstein's smooth spacetime and quantum mechanics' insistence that information can never be destroyed. In Hawking's words:

> 'The absence of event horizons means that there are no black holes, in the sense of regimes from which light can't escape to infinity.'

Hawking also offered a provocative response to the firewall paradox. He suggested that this fiery boundary does not exist at all. In its place, he proposed the apparent horizon, less a point of no return and more a fluctuating veil that temporarily traps matter and information. It is a subtle shift, but a profound one, forcing physicists to rethink what a black hole actually is. In this view, black holes are not eternal vaults. They delay, scramble, and distort information, but eventually release it back into the universe. The proposal is elegant but still untested.

If all this feels confusing, you are in good company. Theoretical physicist Matt Strassler has described the field as a cacophony of competing ideas, with no single framework yet capable of resolving all the paradoxes. Part of the difficulty lies in the mathematics. The effects involved are so delicate that even the most powerful supercomputers cannot reliably track them. Even knowing which questions to ask turns out to be a challenge. As theoretical physicist Joseph Polchinski cautioned:

> 'We really have this conflict between two theories, quantum mechanics and Einstein's theory of gravity, both of which we have strong reason to believe are true.'

The core problem here is that when the two great pillars of modern physics are combined, they appear to predict the loss of information, something most physicists believe cannot be right. For more than four decades, the field has wrestled with this contradiction. Along the way, it has produced ingenious proposals, yet it has also confronted us with even bigger questions and deeper uncertainties.

For philosophers, this is a metaphysical provocation. For physicists, it is a tantalising path forward. Experiments are still catching up, but progress is real. The detection of gravitational waves, ripples in spacetime produced by colliding black holes, has given us new ways to listen to the universe, like stethoscopes pressed against its deepest structures. Future observations of the cosmic microwave background, the universe's earliest light, may reveal subtle fingerprints of space's granular nature or even signs of a holographic origin.

More daring still are experiments that aim to test whether gravity itself is quantum. Researchers are exploring whether quantum superpositions can become gravitationally entangled. If successful, such experiments would suggest that gravity is not merely compatible with quantum mechanics but part of it. Some laboratories are preparing to levitate tiny objects using lasers and probe their gravitational interactions, a scenario that feels like Schrödinger's cat meeting Newton's apple, suspended in midair. And for all their differences, string theory, loop quantum gravity, causal sets, and holographic models share a common ambition: to reconcile the smooth elegance of Einstein's spacetime with the restless, probabilistic world revealed at the Planck scale. This search is not only about equations or curvature. It is about what we mean when we say 'here,' 'now,' or 'real.'

We may not yet have the final answer. But the search itself is reshaping the question. In the coming decades, as quantum computers and space-based detectors come online, we may finally glimpse what lies behind the veil, only to discover that what we took for the screen was never the show at all.

'The more I think about language, the more it amazes me that people ever understand each other at all.'

— **Kurt Gödel**

chapter 4

limits of logic and mathematics

On a crisp autumn day in 1930, a young mathematician named Kurt Gödel, as enigmatic as the problems he pursued, stepped onto a modest stage in Königsberg, Germany. At just twenty-four, Gödel was not yet the star of the gathering. The room itself carried the drama. It buzzed with intellectual electricity and was filled with giants such as David Hilbert, the confident architect of mathematical certainty, and John von Neumann, soon to become a towering figure in mathematics and computing.

The tension in the room made sense once one understood what was at stake. At the heart of the gathering lay Hilbert's dream, sweeping in its ambition. He envisioned mathematics as a perfectly ordered kingdom in which every true statement could be proven from a finite set of axioms, much like following a flawless recipe that never fails. It was the ultimate triumph of human reason: complete, consistent, and unassailable.

Gödel dismantled this vision with devastating elegance. His incompleteness theorems showed that any formal system powerful enough to include basic arithmetic would inevitably contain true statements that could not be proven within the system itself. No matter how carefully constructed, something would always escape. It was like discovering that an infinitely complex puzzle must forever hide a few missing pieces.

This was not a small crack in the foundation. It was a fundamental limit . . . a hard boundary placed on the pursuit of absolute certainty. One might imagine Gödel slipping a quiet note into mathematics' grand instruction manual: the system cannot fully explain itself. For mathematicians who had long trusted logic as an unchanging rulebook, this was as unsettling as learning that the rules of chess might shift mid-game.

Gödel's second theorem pushed the blow even further. Not only could a system not prove all truths within itself, but it also could not prove that it was free of contradiction. The effect of this result was profound. It was not merely a technical setback, but an intellectual earthquake that shattered the belief in mathematics as an eternally stable foundation.

What Gödel quietly presented that day sent shockwaves through the foundations of logic and forever altered our understanding of what mathematics, and truth itself, could hope to achieve. And while it took years for the full implications to ripple outward, the revolutionary significance was immediately apparent: Gödel's work was not confined to mathematics alone but revealed limits that applied to any formal system, whether mathematical, mechanical, or computational.

When Alan Turing later confronted the halting problem—the question of whether a machine can determine if any given program will eventually stop or run forever—he reached a conclusion that echoed Gödel's results. Some problems, Turing showed, are fundamentally unsolvable by any algorithm, regardless of computational power. This was not a technical inconvenience but marked a deep boundary. Together, Gödel and Turing drew a line separating what can be computed, proven, or decided from what must remain undecidable.

For the young field of artificial intelligence, this was a sobering insight. No matter how sophisticated a machine becomes, there will always be questions it cannot answer and patterns it cannot fully resolve. Gödel's quiet revolution thus placed limits not only on human knowledge but also on the very idea of machine intelligence, exposing boundaries that even machines could never cross, long before computers became ubiquitous.

However, the consequences of Gödel's theorems did not stop at the edges of logic or computation. It reached into something profoundly human. His theorems forced us to confront ancient philosophical questions: How much can we truly know? Can reality ever be captured in a complete and final map, or will there always be blind spots no matter how far our theories extend? Immanuel Kant had earlier argued that the mind could never fully access the 'thing-in-itself,' the ultimate nature of reality. Gödel added a startling refinement: even pure reason, when it turns inward, encounters an insurmountable boundary.

This insight was not abstract for Gödel. His own life mirrored the intensity of this search. Deeply introspective and often anxious, he was driven by an almost obsessive need to understand the nature of truth and existence. His close friendship with Albert Einstein underscores this dimension of his character. Einstein once remarked that meeting Gödel felt like encountering a 'mathematical Buddha.' His incompleteness theorems, rigorous though they were, embodied a central human tension: the pull between certainty and mystery, between knowing and not knowing.

Gödel himself described these self-referential limits as a kind of logical hallucination. When a system tries to fully capture and prove everything about itself, it resembles staring too long into a funhouse mirror. The reflection distorts, twists, and refuses to settle into clarity. This distortion is not a flaw but a feature of complexity itself. Moreover, what makes Gödel's insight especially striking is that he did not view this boundary as a failure. Instead, he saw it as a source of creative tension. A system's inability to close the loop perfectly is precisely what keeps it open to new truths, new discoveries, and deeper understanding. In this sense, incompleteness is not a dead end but a form of intellectual breathing room.

There is a fitting historical irony here. Gödel was deeply influenced by Leibniz, the polymath who dreamt of a universal calculus capable of resolving all disputes through pure calculation. Gödel's theorems quietly dismantled that dream, yet they also honoured its spirit. They showed that the pursuit of ultimate clarity is inseparable from the limits we encounter along the way. The boundary itself becomes part of the story, reminding us that reason's power lies not in total closure, but in its capacity to remain open.

In this light, Gödel's theorems are more than milestones in mathematics and logic. They are philosophical signposts, pointing towards a timeless truth: reason has limits, and it is precisely beyond those limits that the richest questions continue to live. And today, as we grapple with the promises and dangers of artificial intelligence, quantum computing, and an unprecedented explosion of data, Gödel's legacy feels newly urgent. He reminds us that some truths will always lie beyond the reach of even the most powerful formal systems. Knowledge, then, is not a journey towards final closure, but an ongoing engagement with what cannot be fully captured.

The Dream of Absolute Certainty

The seismic impact of Kurt Gödel's intellectual revolution cannot be understood without appreciating the context in which it emerged. To see why his work struck so deeply, we must rewind to the early twentieth century, a time when the dream of absolute certainty still shone like a guiding star. And in this era, mathematics was not merely a discipline but regarded as the ultimate fortress of reason, a citadel of eternal truths believed to be immune to history, sensory illusion, or philosophical doubt. Yet even as this confidence reigned, cracks had begun to appear in the foundations.

The nineteenth century had already delivered a series of blows. Non-Euclidean geometry shattered the sacred assumption that Euclid's postulates described the only possible structure of space. Suddenly, alternative geometries emerged that were internally consistent, deeply counterintuitive, and undeniably valid.

Then came Georg Cantor, the maverick who dared to confront the infinite head-on. His theory of infinite sets was as revolutionary as it was unsettling. Cantor's diagonal argument demonstrated that some infinities are larger than others, a result so abstract that it made even seasoned mathematicians uneasy. The backlash to this result was fierce. Leopold Kronecker, a staunch traditionalist, famously declared, 'God made the integers; all else is the work of man.' For critics like Kronecker, the infinite was a Pandora's box best left sealed.

As if this were not destabilising enough, a fault line was exposed at the heart of logic itself. A paradox involving 'the set of all sets that do not contain themselves' revealed a vicious loop. If such a set contains itself, it violates its own definition; if it does not, then it must contain itself. The paradox admitted no stable resolution. It shook the foundations of logic, showing that even the most basic assumptions could collapse into contradiction.

Amid this growing sense of crisis, David Hilbert stepped forward with a grand vision. Hilbert was not merely a mathematician but an architect of certainty. His ambition, later known as Hilbert's program, was nothing less than the reconstruction of mathematics on unshakable foundations. He proposed a fully formal system, an elegant game of symbols governed by precise rules, in which every mathematical truth could be derived mechanically, without intuition or ambiguity. So, if mathematics were imagined as a vast chessboard where every move followed inexorably from the rules, this was Hilbert's ideal. His famous rallying cry, 'Wir müssen wissen; wir werden wissen!' (We must know; we will know), rang out at the 1900 International Congress of Mathematicians in Paris like a declaration of faith. It captured the era's near-religious confidence in human reason, the belief that knowledge could be fully encoded, mechanised, and made certain.

Hilbert set two towering goals. The first was consistency: to prove that the system could never produce contradictions. The second was completeness: to show that every true mathematical statement could, in principle, be proven within the system. At the time, both aims seemed within reach. Mathematics was advancing rapidly in topology, algebra, and analysis. Beyond mathematics, the success of formal logic inspired even grander hopes. Some believed that all knowledge might eventually be reduced to a precise symbolic language, like a cosmic instruction manual waiting to be decoded.

What is less often acknowledged is that Hilbert's program was not merely a technical project. It was a cultural crusade. Early twentieth-century Europe was a continent roiled by political unrest, ideological fracture, and the lingering trauma of World War I. Against this backdrop, Hilbert imagined mathematics as a beacon of universal order, a language so precise and impersonal that it could rise above nationalism and factionalism. In his vision, mathematics functioned almost like an intellectual Esperanto, that is, a neutral, shared framework through which reason itself could restore clarity to a turbulent world. This was not just number crunching. It was an ethical and philosophical wager on the power of rational structure.

At roughly the same time, Alfred North Whitehead and Bertrand Russell were engaged in their own monumental effort to secure mathematics on impeccable logical foundations. Their *Principia Mathematica* pursued rigour with almost ascetic devotion. To prove a statement as simple as '1 + 1 = 2,' they required hundreds of pages of painstaking symbolic derivation. One way to think about this is to imagine mathematics as a film, with this as the exhaustive behind-the-scenes footage, rendering every pixel by hand. Whitehead and Russell wrestled endlessly with primitive notions such as proposition, class, and relation, treating them like untamed creatures that had to be disciplined before any structure could stand.

It is also worth noting that Russell's ambition went far beyond solving isolated puzzles. He sought a logically perfect language in which every sentence mapped cleanly onto a fact, with no room for ambiguity, metaphor, or emotional residue. This was not an ivory-tower exercise. Russell believed that the same vagueness corrupting language also poisoned politics and ethics. Clear logic, he thought, could act as a solvent for social confusion. His broader social vision reflected this faith in rational order, as when he famously remarked, 'The road to happiness and prosperity lies in an organised diminution of work,' imagining a society that functioned with the efficiency of a well-designed system.

However, not everyone shared this outlook. An older philosophical tradition, Platonism, pushed back forcefully. Where Hilbert and Russell treated mathematics as something constructed by human minds, Platonists such as G. H. Hardy regarded it as an eternal realm waiting to be discovered. In his quietly elegiac

A Mathematician's Apology, Hardy described mathematicians as artists who do not invent patterns but reveal them, uncovering the hidden architecture of a timeless mathematical universe. To Hardy, showing that a number like 317 is prime was not a creative act but a moment of recognition, a brief glimpse of an objective structure that existed independently of human thought.

This reverence extended beyond well-known figures. Kurt Hensel, the pioneer of p-adic numbers, worked in a domain that many contemporaries considered bizarre, even mystical. P-adic numbers, now indispensable in number theory, were once viewed as arcane curiosities. But for Hensel and fellow Platonists, such entities were no less real than physical objects. Their strangeness was not a defect but a sign that mathematics reaches into depths as mysterious as those explored by physics or metaphysics. Seen in this light, Hensel's work was not an eccentric detour but an expression of a broader philosophical faith.

This faith, shared in different forms across the mathematical world, soon crystallised into a powerful consensus about what mathematics ought to be. By the late 1920s, the pursuit of absolute certainty in mathematics had come to rest on three formidable pillars. Hilbert's formalism promised a flawless, self-contained symbolic system. Russell's logical atomism aimed to produce a perfectly transparent logical map of reality. Hardy's Platonism insisted that mathematical truths were timeless and objective, waiting patiently to be uncovered. Together, these approaches formed a dazzling intellectual cathedral, one that seemed capable of capturing the full power of human reason and banishing uncertainty once and for all.

What no one yet fully grasped was that the very tools that made this project so compelling, formal precision and self-reference, would also expose its deepest vulnerability. Kurt Gödel completely demolished this grand construction by revealing that absolute certainty was not a summit waiting to be conquered, but a mirage that recedes as one approaches it. With this insight, the ground beneath mathematics shifted, and with it, our understanding of truth itself.

Gödel and the Architecture of Self-Reference

So far, we have seen how Kurt Gödel's incompleteness theorems struck at the towering ambition of mathematical absolutism with devastating force. Gödel's genius lay in making mathematics turn inward and speak about itself. Like a mirror attempting to capture its own frame, the move was subtle yet explosive, permanently reshaping the landscape of logic.

Yet this kind of self-reference was not entirely new; it had deep philosophical ancestors. Gödel's insight can be viewed as a mathematical descendant of ancient philosophical puzzles that had troubled thinkers for millennia, most famously the Liar Paradox attributed to the Cretan Epimenides. His declaration, 'All Cretans are liars,' creates a self-entangling knot. If the statement is true, it is false; if it is false, truth slips away again. Later logicians, such as Eubulides, refined this into the spare and unsettling sentence, 'This sentence is false,' a riddle that toys with meaning and truth like a philosophical Rubik's Cube. This idea was also taken up by thinkers such as Leopold Löwenheim and Thoralf Skolem, who had already glimpsed the unsettling possibility that some truths might lie forever beyond formal reach. Therefore, what had long been dismissed as a linguistic curiosity became Gödel's secret weapon. Self-reference, he showed, was not a harmless trick but a doorway to deep and unavoidable limitations.

To turn this philosophical understanding into a mathematical instrument, Gödel needed a way to make self-reference precise rather than poetic. His masterstroke was the invention of what we now call Gödel numbering, a technique that is as elegant as it is quietly subversive. He assigned each symbol in a formal system a unique natural number, translating statements and proofs into numerical sequences. In effect, mathematics learnt to encode its own language and reason about itself from within. This allowed formal systems to reflect not just on what they assert but on how they make those assertions in the first place.

The first stepping stone in this direction was Gödel's first incompleteness theorem. At its heart lies a startling claim: in any consistent formal system powerful enough to express basic arithmetic, there exist true statements that cannot be proven within the system itself. The logic is almost theatrical. Gödel constructs a statement that effectively says, 'This statement cannot be proven here.' If the system could prove it, the system would contradict itself. If it cannot, then the statement is true but unprovable. Either way, the dream of capturing all mathematical truth within a single formal net collapses. Completeness, once thought attainable, is revealed as an illusion.

Gödel's second incompleteness theorem delivers the final blow, which states that no such system can prove its own consistency. In other words, a system can never fully assure itself that it will not generate contradictions. Hilbert's rallying cry, 'We must know; we will know!' ran headlong into a fundamental barrier. Gödel demonstrated that certainty has intrinsic limits, and that some truths remain permanently beyond formal validation.

Moreover, the recursive pattern at the heart of Gödel's work does not remain confined to logic and mathematics. It extends naturally into language, where it underwrites human meaning-making itself. In the mid-twentieth century, Noam Chomsky transformed linguistics by formalising grammars that incorporate recursion. Context-free grammars, in particular, allow rules to refer to themselves, making it possible to generate theoretically infinite structures from finite means. Consider the sentence: 'The rat the cat the dog chased caught ate the cheese.' Clauses nest within clauses like Russian dolls, producing unbounded complexity from a limited set of elements. Recursion, then, is not a quirk of

English but a universal feature of human language. It enables us to express infinitely many thoughts with a finite vocabulary, just as Gödel showed that formal systems can encode infinite chains of reasoning within finite symbols.

Douglas Hofstadter brought this insight into cultural focus in *Gödel, Escher, Bach*. He coined the term 'strange loops' to describe systems that seem to ascend through levels of abstraction only to circle back to where they began. Escher's drawing of two hands sketching each other offers a striking visual analogue. Such loops capture the logic of self-reference in art, music, and cognition—the very phenomenon Gödel revealed at the core of formal systems.

Yet Gödel's discoveries did not bring mathematics to a halt. In their wake, mathematicians did not abandon rigour; they refined it. One response emerged in the form of reflection principles, which allow a formal system, under carefully specified conditions, to assert that if it proves a statement, then that statement is true. Here, reflection functions like a mirror. Meaning, the system examines limited aspects of its own structure without collapsing into contradiction. It is self-referential, disciplined, and contained.

Solomon Feferman, a quieter but formidable figure in mathematical logic, carried this development further. Rather than treating Gödel's incompleteness theorems as a final barrier, he read them as an invitation to continue. He called this idea reflective closure, and it can be illustrated with the following example.

Imagine a hall of mirrors in which each reflection can assess the one before it and then gesture towards another beyond. In Feferman's framework, each new mathematical system can look back and affirm the consistency of its predecessor, creating a sequence of ever-stronger theories. There is no ultimate, all-encompassing system as Gödel's second incompleteness theorem rules that out. But the absence of final closure does not end the ascent. It transforms limitation into propulsion. What once appeared to be the edge of the map begins to look like an opening into unexplored terrain.

What makes this idea especially intriguing is how naturally it resonates with debates about consciousness. Philosophers and cognitive scientists have long wrestled with a central puzzle: how does a finite, biological organ generate the rich, self-aware experience of being a self? One influential proposal holds that consciousness arises through internal modelling—a system capable of representing itself from within. In this light, the mind begins to resemble a 'strange loop,' a structure that turns back upon itself in ever more refined acts of self-reference.

Seen in this light, the parallel to Gödel becomes hard to ignore. Just as a formal system cannot step outside itself to certify its own completeness, the mind cannot attain a perfectly transparent account of itself. There is no final vantage point from which the whole of consciousness can be surveyed without remainder. Yet this limitation does not spell paralysis. The mind can reflect, revise, and deepen its self-understanding without end. Therefore, self-awareness, like mathematical reflection, is not a finished achievement but an ongoing movement.

It is precisely this shared structure of reflection that invites a broader comparison. Here, the boundaries between disciplines begin to dissolve. On one side stand the austere symbols and axioms of logic; on the other, the elusive texture of subjective experience. Yet both domains are shaped by the same underlying patterns of recursion, self-reference, and fixed points. Even mathematics carries an intuitive image. The Brouwer fixed-point theorem, quietly foundational to many of these ideas, can be expressed in homely terms: stir a cup of coffee, and at least one drop must end up exactly where it began. However turbulent the motion, some point remains unchanged. It seems that loops and returns are woven into the structure of space itself. Therefore, self-reference is not a defect to be eliminated but a generative feature of complex systems. In logic, computation, language, and art, recursion marks the moment when structure folds back upon itself and depth emerges. It is finitude touching infinity, a fold in the fabric of knowledge that reveals new layers each time it is traced.

From the unmoving drop in a stirred cup to the infinite nesting of human language, from Gödel's unprovable truths to Escher's impossible staircases, fixed points and recursion form a hidden rhythm beneath thought itself. What Feferman contributes to this pattern is a philosophy of continuation. Gödel's theorems are not a brick wall but a threshold. Each act of reflection gives rise to a stronger framework; each system invites another to extend it. There is no final, all-encompassing summit, yet the ascent is not futile. It resembles a mythic climb—not the endless frustration of Sisyphus, but an ever-renewed rising in which every step expands the horizon.

In this light, Gödel's theorems are not technical curiosities confined to formal logic. They are keys to a labyrinth connecting mathematics, mind, language, and art. They draw us into an infinite hall of mirrors where every reflection generates another, and where understanding grows precisely by turning back upon itself. In that looping movement, self and world intertwine, revealing a quiet truth: to know is to reflect, and to reflect is to become continually new.

An interesting fact here is that Gödel completed his groundbreaking proof in 1930, but in keeping with his reserved and enigmatic nature, initially kept it to himself. He finally revealed it that September at a conference in Königsberg, the very city where, nearly a century earlier, Immanuel Kant had famously grappled with the limits of human reason. The irony is hard to miss. Kant had approached those limits philosophically, asking what reason could and could not achieve. Gödel arrived with a definitive mathematical demonstration that some questions are, in principle, beyond reason's power to settle.

His full results appeared in 1931 in a dense but now legendary paper titled *Über formal unentscheidbare Sätze der Principia Mathematica und verwandter Systeme* (*On Formally Undecidable Propositions of Principia Mathematica and Related Systems*), a title formidable enough to deter all but the most determined readers. Yet its impact was immediate among those who grasped its implications. John von Neumann, himself a towering figure of twentieth-century science and a polymath at home in mathematics, physics, and early computing, recognised the magnitude of the discovery at once. After hearing a summary, he reportedly turned to Gödel and said simply, 'This changes everything.' He also concluded that Hilbert's grand program was effectively over, as if the final curtain had fallen on a long-running intellectual drama.

There is something almost poetic in this outcome. The quest for absolute certainty, driven by the optimism of thinkers like Hilbert, did not uncover an impregnable fortress of truth. Instead, it revealed a labyrinth filled with unsolvable problems and unexplored corridors. And just because Gödel exposed the limits of formal systems, it does not follow that he was a nihilist. Beneath his austere logic lay a deeply held belief in the objective reality of mathematical truth. A committed Platonist himself, he regarded mathematical entities as timeless and independent of human thought, with formal systems serving only as imperfect windows into that higher realm. Yet the practical significance of Gödel's constructions took decades to manifest in concrete mathematics.

For many years, critics dismissed Gödel's self-referential constructions as clever curiosities with little relevance to everyday mathematics. This scepticism was decisively challenged in 1977 by Jeff Paris and Leo Harrington, who uncovered a striking phenomenon now known as natural undecidability. They began with Ramsey's theorem, a familiar result in combinatorics stating that in any sufficiently large system, some form of order is guaranteed to emerge, no matter how chaotic the arrangement may appear. Paris and Harrington then introduced a subtle but powerful modification. They required that the guaranteed orderly subset not only exist but also be at least as large as its smallest element. This seemingly innocuous condition transformed the theorem completely. The resulting statement is unquestionably true, yet it cannot be proven within Peano arithmetic, the standard axiomatic system for the natural numbers. What looked like a minor adjustment exposed a profound limitation of formal reasoning.

Thus, it turned out that undecidability was not confined to the abstract edges of logic. It was already present in the familiar territory of finite mathematics. The Paris–Harrington theorem served as a wake-up call, showing that undecidable truths can arise naturally in ordinary mathematical practice. Since then, many similar examples have been found across fields such as graph theory and topology, confirming that Gödel's influence extends to nearly every corner of mathematics.

It is also important to point out that Gödel's vision was never limited to theorems and proofs alone; it extended to the deepest questions of existence. In a candid letter to his mother, Gödel even argued that an afterlife must exist, or else life would be stripped of meaning altogether. For him, logic, metaphysics, and the search for ultimate truth were inseparable. But here lies a paradox worthy of Gödel himself. His theorems humbled human reason by exposing its limits, yet they also elevated it by pointing towards a reality that transcends formal proof. Our systems may falter and our axioms may fall short, but the realm of truth they gesture towards appears vast and inexhaustible, forever inviting exploration beyond the borders of certainty.

Gödel and the Limits of Proof

We have explored Gödel's incompleteness theorems and self-referential constructions, but the deepest legacy of his work lies in the concept of independence, which states that certain mathematical statements are so fundamental that they can neither be proved nor disproved from a given set of axioms. The most famous example of this is the continuum hypothesis, proposed by Georg Cantor in the late nineteenth century. It asks whether there exists a size of infinity strictly between that of the integers and that of the real numbers.

In 1940, Gödel showed that the continuum hypothesis cannot be disproved using the standard axioms of set theory, known as Zermelo–Fraenkel set theory with the axiom of choice. The axiom of choice, often seen as both indispensable and controversial, allows one to select an element from each set in an infinite collection even when no explicit rule for making those selections is available. It enables powerful results such as Tychonoff's theorem, while also giving rise to counterintuitive consequences like the Banach–Tarski paradox, in which a sphere can be decomposed and reassembled into two identical copies of itself.

More than two decades later, in 1963, Paul Cohen introduced the method of forcing and proved the complementary result: the continuum hypothesis cannot be proven from those same axioms either. As a result, the hypothesis was left suspended in a liminal state, neither true nor false within the accepted framework of set theory.

This discovery fundamentally altered the notion of mathematical truth. The seemingly absolute realm of mathematics revealed itself as far more intricate and mysterious than previously imagined, filled with questions whose answers depend on the axioms we choose to accept. In confronting these limits, mathematics does not collapse into uncertainty but instead invites a deeper reflection on what it truly means to know.

In summary, the continuum hypothesis was not false, but it was not true either. It was something stranger: independent. It could neither be proved nor disproved using the standard axioms of set theory. This is similar to stories with alternate endings. One could either accept or reject the continuum hypothesis, and in either case, arrive at a coherent, logically consistent universe.

It should also be noted that this was not an isolated incident. Soon, even the axiom of choice, once a polite mathematical scandal, revealed its own uneasy status. Certain formulations of it proved undecidable within the core of Zermelo–Fraenkel set theory. Alongside this came a dazzling array of other exotic propositions. Large cardinal axioms, for instance, posit immense and almost otherworldly infinities that lie far beyond the reach of ordinary mathematical tools. These are not merely very large numbers; they are the Mount Everests and black holes of the mathematical cosmos, shaping the structure of entire universes of thought.

As Saunders Mac Lane, one of the architects of category theory, observed with disarming clarity:

> 'Mathematics is not one thing, but many.'

The important takeaway here is that he meant this statement literally. The vision of mathematics as a single, gleaming tower of Platonic truth gave way to something far more expansive. What emerged was a vast continent of parallel mathematical worlds, each internally consistent, each governed by its own rules. Euclid's orderly geometric garden had transformed into something closer to a jungle, rich, tangled, and alive with possibility.

The idea that mathematics could splinter into multiple, incompatible truths may sound like science fiction, but it now sits firmly within the domain of serious thought. Modern cosmology offers a striking parallel. The multiverse is no longer just a comic-book fantasy; it is a genuine hypothesis in contemporary physics. Perhaps reality itself, whether mathematical or physical, is fundamentally plural, branching outward like the strands of Indra's net, infinite and interlinked. In this way, the mathematics of the twentieth century began to resonate uncannily with the physics of the twenty-first.

Hilbert's old dream of reducing mathematics to a flawless engine of truth, humming with mechanical precision, did not survive this upheaval. What replaced it was no less beautiful, but far more untamed: mathematics as a living exploration of the possible. And Gödel's revolution continues, quietly radical and endlessly generative, pushing the limits of reason even as it gestures, paradoxically, towards something beyond it.

The Expanding Universe of Undecidability

Gödel's incompleteness theorems were not mathematical fireworks that briefly dazzled and faded. Instead, they marked the opening act of a revolution that reshaped logic, mathematics, and eventually the way we think about computation itself. Therefore, what began as a sober insight into the limits of formal systems soon expanded into a far-reaching story of undecidability, randomness, and unexpected complexity hiding inside even the most rigorous intellectual frameworks.

In 1936, this story took a decisive turn with a young British mathematician named Alan Turing. Today, Turing's name evokes modern computers, artificial intelligence, and wartime codebreaking, but at the time, he was grappling with a deceptively simple question that had haunted logicians since David Hilbert posed his *Entscheidungsproblem* (decision problem): Could there exist a purely mechanical, foolproof method—a universal algorithm—that could decide whether any mathematical statement was true or false?

Turing's answer was an unequivocal no. To demonstrate this, he introduced a conceptual device now known as the Turing machine, an elegant and minimalist model that remains the foundation of theoretical computer science. The brilliance of his approach lay in its use of self-reference, echoing Gödel's strategy, by imagining a program designed to analyse its own behaviour. This led to the halting problem, which states that no algorithm can determine, in all cases, whether a given program will eventually stop or run forever. Turing's negative result showed that some problems lie permanently beyond the reach of algorithmic decision-making, regardless of how powerful or sophisticated our machines become. As Scott Aaronson has dryly noted, undecidability results 'typically rely on self-reference or something equivalent to it.' Far from being a loophole or a nuisance, self-reference marks the edge of what systems can know about themselves.

It is important to recognise that this insight carries real-world consequences. It means there can never be a perfect, all-knowing software verifier. Even as you read this on a phone or computer, deep within the device are processes whose behaviour can never be completely guaranteed. Whether the code controls a Mars rover or regulates a pacemaker, a residue of uncertainty is built into the logic itself. The vision of flawless digital reliability remains, in principle, unattainable.

However, there is a curious twist in this story. Turing himself played a central role in building some of the earliest practical computers during World War II at Bletchley Park, where he helped crack Nazi codes. He knew firsthand that machines could achieve astonishing feats, yet he also understood that their power was forever bounded by theoretical limits embedded in computation.

Several decades later, in the 1970s, Gregory Chaitin carried these ideas further. Drawing on the legacies of Gödel and Turing, Chaitin developed algorithmic information theory and introduced a startling concept: algorithmic randomness. His work suggested that some mathematical truths are not merely unprovable but fundamentally patternless, resembling noise rather than structure. Even mathematics, it seemed, contained regions that resist compression, prediction, and tidy explanation.

To grasp the full radicality of his insight, one can focus on a single, enigmatic number that embodies this unpredictability. At the heart of Chaitin's work lies a particularly enigmatic object known as Omega (Ω), the halting probability. Imagine generating a computer program at random by flipping coins to choose each instruction, then asking what the probability is that this program will eventually halt. Omega is that probability, a single real number that encapsulates the ultimate unpredictability of computation.

What is particularly striking is that Omega's digits are maximally random. Although the number is precisely defined, no algorithm can compute more than a finite initial segment of its expansion. Each additional bit encodes the answer to an undecidable question about whether some program halts. Therefore, omega can be viewed as an infinite library of unsolvable problems compressed into a single sequence of numbers.

What makes Omega so captivating is the way it overturns our earlier picture of incompleteness. Gödel's theorems had suggested that undecidable statements were rare and exotic, hidden deep in the formal forest like strange anomalies among otherwise orderly trees. Chaitin revealed something far more radical: a mathematical landscape in which incompleteness is everywhere, a dense jungle where randomness pervades even the most pristine regions of pure logic.

Philosophically, Chaitin saw his ideas reflected in earlier thinkers. He found affinities with Leibniz, who wrestled with the tension between simplicity and complexity centuries ago, and with John von Neumann, who viewed randomness as a fundamental resource in both mathematics and physics. Omega, in this sense, echoes the spirit of Heisenberg's uncertainty principle, not in the physical world but in the structure of mathematical truth itself. It stands as a stark reminder that unpredictability is woven into the very fabric of logic.

Ultimately, from Gödel's incompleteness theorems to Turing's halting problem and onward to Chaitin's randomness, the twentieth century revealed a new intellectual landscape where certainty became partial, knowledge provisional, and mystery inseparable from truth.

Beyond Classical Frameworks

For centuries, classical logic wore the crown of certainty, laying down the rules of rational thought much as Euclid did for geometry. Aristotle sketched the first lines, and thinkers from Spinoza to Russell dutifully stayed within them. Logic was not merely 'a' tool; it was 'the' tool. The philosopher's scalpel, the scientist's compass, the mathematician's anchor.

Then quantum mechanics arrived, and certainty cracked. Unlike many intellectual revolutions, this one did not emerge from a philosopher's study but from the clicks of Geiger counters and the strange behaviour of particles in the atom's hidden theatre. Reality refused to sit still. Electrons were not 'here' or 'there' but somehow both, until observation forced a decision. It was as if Schrödinger's cat, upon being watched, chose whether to purr or perish.

Confronted with this unsettling world, John von Neumann and Garrett Birkhoff proposed a startling possibility: perhaps the problem was not reality, but the logic we used to describe it. What if the venerable distributive law of classical logic, the rule that insists A and (B or C) must equal (A and B) or (A and C), simply did not hold in the quantum domain? What if reality, at its deepest level, behaved less like a tidy syllogism and more like a Zen koan?

From this question emerged quantum logic, a radical rethinking of truth itself. In this framework, measurement does not merely reveal reality; it helps bring reality into being. Put differently, asking a question changes what can be answered. Therefore, it becomes evident that in quantum logic, propositions behave in relational, context-sensitive ways and depend on order.

With further radical developments across the twentieth and twenty-first centuries, the belief in a single, universal logic underwent a profound transformation. Modal logics flourished, offering ways to reason about necessity, possibility, obligation, and even time itself. Temporal logic, for instance, allows us to speak rigorously about change, futures, and branching timelines. Non-monotonic logic abandoned the assumption that knowledge must only grow. In these systems, revising your beliefs in light of new information is not a failure of reason but a requirement of it.

Then came fuzzy logic. Introduced by Lotfi Zadeh in 1965, it challenged the idea that truth must be all or nothing. What this means is that instead of sharp boundaries, it allowed degrees of truth, closer to how humans actually think. This is the logic of weather forecasts, moral judgements, and inherited recipes. It also happens to run air conditioners, control systems, and many forms of artificial intelligence.

Beyond these came a series of unsettling discoveries that showed that not only are our truths incomplete, but the very frameworks through which we reason may not be unique. In other words—logic itself, long treated as the most stable of intellectual compasses—began to fracture into multiple, equally legitimate paths. One of the major thinkers to play a crucial role in this shift was L. E. J. Brouwer. The Dutch mathematician refused to see logic as a divine code written into the fabric of the universe. Instead, he treated it as something intimate and human, rooted in mental activity. For Brouwer, mathematics was not a cold Platonic temple, but a structure patiently assembled in the mind, step by step.

He led a movement known as intuitionism, which rejected one of classical logic's oldest rules: the law of the excluded middle. According to this principle, every mathematical statement must be either true or false, even when we have no way of determining which. Brouwer challenged this view, arguing that unless a statement can be explicitly proved, we have no grounds to call it true. In the absence of proof, it remains suspended—neither affirmed nor denied—lingering in a kind of mathematical limbo, conceivable in thought but not yet brought into existence.

This was no minor disagreement. In Brouwer's universe, infinite sets lost their status as completed, eternal objects. They became processes rather than monuments. To put it another way, an infinite set was not something you could survey all at once; it was more like a staircase that never ends. You can always take another step, but you will never stand at the top. Mathematics, in this view, was not about being but about becoming.

The roots of this philosophy ran deep. They take us back to a famous line by Leopold Kronecker: 'God made the integers; all else is the work of man.' Brouwer, however, went even further. Even the integers, he argued, were not divine gifts but constructions of the human mind. They existed only through acts of human construction, brought into being by the mind rather than discovered in a heavenly realm.

While Brouwer offered a logic of restraint and careful construction, another movement arose that looked contradictions squarely in the face and refused to look away. This was the birth of paraconsistent logic. Let's take a closer look.

In classical reasoning, contradictions are catastrophic. Accept one, and the system collapses; from a single inconsistency, anything follows; true and false blur into nonsense; arithmetic breaks; and reason itself implodes. But paraconsistent logics, developed by the Brazilian logician Newton da Costa and later extended by thinkers such as Graham Priest, take a different approach. In these systems, contradictions do not explode. They are contained and smoulder without consuming everything around them.

Priest pushed this idea further with dialetheism, the provocative claim that some contradictions are genuinely true, both true and false at the same time. It sounds outrageous, but it mirrors how many real systems actually work. In computer science, large databases routinely contain conflicting information, yet we do not discard the entire system; we manage the inconsistency. And once we stop treating contradiction as an automatic failure, its presence elsewhere becomes impossible to ignore.

Moreover, this tolerance for inconsistency is not confined to machines or formal theories. In law, rules often pull in opposing directions while the legal framework continues to function. And in ordinary life, we carry contradictions effortlessly: we love people we resent, cherish ideals we routinely fail to uphold. Human cognition, it turns out, is quietly paraconsistent.

All of this points to a dizzying conclusion. There are not only different truths, but different ways of reasoning itself. Different logics. Different mathematical worlds. Like rival mythologies or parallel universes in speculative fiction, these systems do not cancel one another out; they coexist.

Taken together, these developments signal a profound shift. Where logic once appeared as a single imperial authority, rigid and eternal, we now find ourselves in a pluralistic archipelago of logics. Some are strict and formal, while others are fluid and accommodating. Some wear tuxedos, while others prefer tie-dye. As the philosopher Susan Haack observed, classical logic remains a great lighthouse. But it now shines alongside other beacons, illuminating not one path but many.

Therefore, we can say that each logic acts as a lens, revealing different aspects of a richly faceted reality. We no longer speak of 'the' truth, but of 'a' truth. We no longer ask, 'Is this logical?' but instead, 'Which logic applies here?' And once logic is seen as plural and situated, a deeper question quietly emerges about the nature of reason itself. Perhaps logic, much like geometry, music, or morality, is inherently contextual. Perhaps what we call rationality is simply our best attempt to move in step with the rhythms of the universe. And perhaps reason itself is not a rigid lawbook after all, but something closer to poetry. A fascinating parallel emerges here. Just as paraconsistent logics allow contradictions without collapsing into chaos, long before these developments in the West, Indian philosophical traditions such as Nyāya and Jain Syādvāda had already embraced multivalent and contextual notions of truth. In these systems, 'maybe' is not evasive; it is principled.

Seen from this broader vantage point, we have entered the age of logical plurality, where mathematics begins to resemble less an ivory tower and more a bustling bazaar of ideas. Some regions are carefully constructed by cautious intuitionists. Others shimmer with contradictions held in delicate balance. And running through them all is a quietly exhilarating reminder: even in the most precise and rigorous of disciplines, mystery endures. In this expanding cosmos of reason, there is room for the ascetic, the rebel, the mystic, and even the paradox.

The Silence between Axioms

In 1931, a quiet storm broke over the neat landscape of mathematics, a storm not of numbers but of negation. Kurt Gödel, as noted earlier, had dismantled the formalist hopes of Hilbert and his followers, yet he himself remained an unapologetic Platonist. He believed that mathematical truths exist in an eternal, unchanging realm beyond space and time, closer to Olympus than to algorithm. He once wrote to his mother—a devout woman puzzled by his philosophical leanings—that we perceive numbers and sets not with our eyes but through a form of rational intuition, something like an inward-turned third eye. It is striking that the very man who fractured the foundations of mathematics would describe his worldview in such almost spiritual terms.

Moreover, Gödel did not see the incompleteness theorems as a defeat of this belief. On the contrary, he saw them as revealing something profound: truth outruns proof, and formal systems are too cramped to contain all of it. 'Either mathematics is too big for the human mind,' he wrote, 'or the human mind is more than a machine.' That single sentence ignited decades of debate and gave rise to what later came to be called, with some grandiosity, optimistic Platonism. And once that possibility was placed on the table, others began to ask how such a mind might work and where its powers could possibly come from.

Among those who took up this challenge was Sir Roger Penrose, a knight errant of mathematical metaphysics. In *The Emperor's New Mind* and *Shadows of the Mind*, Penrose dared to link our ability to grasp mathematical truth with the deepest mysteries of quantum physics. Consciousness, he suggested, might not arise merely from neurons firing like circuits, but from unknown laws operating at the Planck scale, where spacetime itself flickers in superposition. It was a bold synthesis of neuroscience, metaphysics, and quantum drama. Penrose even proposed that within the microtubules of brain cells, yes, microtubules, lay the quantum machinery that allows the mind to access truths no algorithm can compute. Sceptics scoffed at this bold claim but for a generation of intellectual romantics, this idea opened a back door to Gödel's Platonic heaven, accessed through the human skull.

And yet the puzzle remains stubborn. If Platonic truths are infinite, perfect, and timeless, how does a finite, heat-sensitive organ like the brain glimpse them at all? Are we tuned to some abstract broadcast beyond time, or merely mistaking pattern for harmony?

While these questions lingered, the formalists regrouped. Hilbert's dream of a self-contained, provably consistent mathematics had sunk, but formalism itself did not die; it evolved. Instead of searching for a single ultimate foundation, mathematicians began constructing many formal systems, treating mathematics as a landscape of possible worlds rather than a single sacred structure.

Harvey Friedman's work in reverse mathematics turned the question inside out. Rather than asking what follows from which axioms, he asked which axioms are truly required to prove the theorems we already value. The answer was often surprising. Many results taught at the undergraduate level require only remarkably weak logical systems. You do not always need the skyscraper of ZFC set theory. Sometimes a well-furnished shack will do. This was more than a technical insight. It marked a philosophical shift, from absolute certainty to local usefulness, from eternal foundations to pragmatic construction.

It is at this juncture that Imre Lakatos entered the story. A Hungarian exile with a taste for dialectics and Socratic drama, Lakatos rejected the sterile 'theorem–lemma–proof' ideal of postwar mathematics. In *Proofs and Refutations*, he argued that mathematics lives not in final certainty but in ongoing argument where proofs evolve, concepts shift, and counterexamples provoke revision. Mathematics, for Lakatos, was not a mausoleum of truths but a living conversation.

He famously said:

> 'Blind commitment to a theory is not an intellectual virtue; it is an intellectual crime.'

Lakatos staged the history of mathematics as a courtroom drama, where every conjecture faces cross-examination, every proof is vulnerable to challenge, and truth emerges not from axioms alone but from struggle, creativity, and social dialogue. Look closely at the record and his point becomes hard to miss. Imaginary numbers were once dismissed as 'sophistical.' Negative numbers were treated as absurd, likened to debts that should not exist. Infinitesimals fared even worse, effectively burned at the philosophical stake until Abraham Robinson resurrected them through non-standard analysis, slipping them past the logical border guards of the twentieth century.

Seen this way, mathematics resembles less the deciphering of ancient runes and more the composition of a vast symphony, its instruments and motifs constantly changing. It is part invention and part discovery, an evolving, co-authored work written not only in symbols but in chalk dust, peer review, dissent, and the occasional flash of inspired madness. Lakatos pushed this view further by questioning the very idea that scientific laws can be derived cleanly from facts. As he put it:

> 'One can today easily demonstrate that there can be no valid derivation of a law of nature from any finite number of facts; yet we still keep reading about scientific theories being proved from facts. Why this stubborn resistance to elementary logic?'

So where does this leave us? Are we merely sketching shadows on the wall of Plato's cave? Are we somehow tapping cosmic truths through microtubules and mathematical reverie? Or are we simply glorified pattern-seeking animals, weaving elegant sense out of the useful illusions evolution has handed us? The answer may lie somewhere in between, or nowhere definite at all.

Still, a further possibility suggests itself. What if mathematical truth is neither sequestered in a distant Platonic heaven nor entirely manufactured on the factory floor of human culture? Between austere Platonism and rebellious constructivism, a quieter and subtler revolution has been taking shape. It is a middle path, not unlike the one the Buddha proposed between indulgence and asceticism, and it has begun to reshape how philosophers think about the mathematical realm.

Structuralism pushes this idea in a different but complementary direction. Echoing the logic of a Bach fugue or the geometry of a Mughal *jali* (screen), philosophers like Stewart Shapiro argue that mathematics is not about objects as things, but about positions within structures. The number 2 is not a chunk of abstract substance; it is a role in a pattern, like the second violin in a quartet. The structure itself enforces discipline and coherence, even if our paths into it, through notation, pedagogy, or intuition, remain deeply human. Put simply, meaning arises not from what a number is in itself, but from how it moves within the pattern. This shift is no minor adjustment. It opens the door to a kind of mathematical multiverse, less Marvel spectacle and more *Mahāyāna* cosmology, where many worlds coexist, each internally rigorous, each illuminating a different facet of the infinite.

The origin of this idea can be traced back to Richard Dedekind in the nineteenth century. For Dedekind, the number 2 is not special because of any internal substance; it is simply the successor of 1 within a carefully defined system. In this light, numbers are positions in a meticulously choreographed structure. Moreover, Dedekind did not announce this as a manifesto, but what he offered was radical: a way of understanding numbers not as self-contained entities, but as roles in a relational order. It was the first whisper of what would later become structuralism's full voice.

Then came Paul Benacerraf, wielding a philosophical scalpel. In his 1965 paper, he cut directly into our assumptions about numerical identity. If the number 3 can be modelled equally well by different set-theoretic constructions, what, exactly, is 3? Benacerraf's insight was not a gentle refinement but a jolt to the foundations. Numbers, he argued, do not come with unique identity cards. As long as an object plays the right role within a structure, it qualifies. Numbers began to look less like artefacts locked in a metaphysical vault and more like actors capable of performing the same role across different productions. With this move, structuralism became not just an option but a necessity.

Building on this shift, Michael Resnik introduced a form of Platonic restraint known as ante rem structuralism. For Resnik, structures such as the natural numbers exist independently of any particular system that instantiates them. Under this interpretation, a number is not a chunk of metaphysical material; it is a position within a network of relations, like a point on an abstract map, and one need not survey the entire terrain to acknowledge the grid. Resnik's approach gave philosophers a way to speak seriously about mathematical truth without becoming entangled in debates over what numbers are made of.

Stewart Shapiro extended this vision with a powerful metaphor. In *Philosophy of Mathematics: Structure and Ontology*, he suggested thinking of mathematics as music. A symphony exists whether or not an orchestra is playing it. It is a structured possibility, waiting to be realised. In the same way, the structure of the natural numbers exists independently of any particular set-theoretic model. The result is a vision of mathematics that feels less like a dusty archive and more like a living score, always ready for performance.

Structuralism, however, did not go unchallenged. Penelope Maddy, characteristically sceptical of metaphysical excess, asked whether we really need to posit these abstract structures at all. Her naturalised Platonism approaches mathematics the way a seasoned traveller approaches a city: not because it appears on a metaphysical map, but because walking its streets has proven coherent, reliable, and fruitful. For Maddy, mathematical objects originate in the messy terrain of human thought, but once embedded in formal theories, they acquire a kind of autonomy. Not because they inhabit an otherworldly realm, but because they are stabilised by the internal constraints of their systems: consistency, elegance, parsimony, and even aesthetic demand. In an age wary of grand metaphysical claims, her position offered a refreshingly pragmatic refrain.

Geoffrey Hellman, by contrast, embraced abstraction through the modal lens. Why claim that numbers simply 'are' when we can say that if certain structures exist, then specific consequences follow? Given this conception, mathematics becomes a universe of conditional truths—a kind of logic theatre that asks for no belief in the pieces, only a commitment to the rules. We suspend disbelief, letting hypothetical worlds play out. No metaphysical baggage, no ontological guilt, just structures as possibilities.

But abstraction need not be timeless or static. It is here that Øystein Linnebo adds a temporal dimension, inspired by potentialism in set theory. Instead of frozen monuments, mathematical structures become ever-unfolding scrolls, revealing themselves gradually as we explore. This mirrors how discovery actually unfolds: layer by layer, insight by insight, as if peeling back meaning itself. In Linnebo's hands, mathematics is not a gallery of statues but a living narrative.

Marcus Giaquinto took a different approach, reminding us that the mind is part of the equation. Structuralism, he argued, must account for human cognition: how we perceive, imagine, and inhabit abstract structures. He focused less on metaphysical commitments than on experience itself. Even the most austere mathematical architecture requires eyes to admire it, hands to trace it, and minds to grasp its elegance.

Thus, from Dedekind's careful successor chains to Giaquinto's mindful mappings, the story of structuralism is a journey from rigid definitions to relational insight. Within these frameworks, numbers cease to be isolated truths and become nodes in a network, actors in a play, notes in a timeless composition. And the lesson here is simple yet profound: trade solidity for structure, and you gain a deeper, more connected form of mathematical beauty.

On the whole, just as physics accommodates the coexistence of classical mechanics, general relativity, and quantum field theory—each valid in its own domain—mathematics too resembles a cathedral rather than a castle, with many wings, each holding its own form of truth. In this cathedral, contradictions need not collapse; they only require contextual clarity. This maturation of reason reveals something profound about ourselves: our finite, fallible minds are nonetheless capable of glimpsing and shaping infinite patterns—not as passive recipients of cosmic telegrams, nor as deluded inventors of fantasy, but as active participants in an ongoing dialogue between logic and life.

The Complexity Chronicles

When we think about what computers can do, it is tempting to picture a digital wizard conjuring answers at lightning speed, solving problems, cracking codes, and forecasting outcomes. Yet beneath this polished surface lies a quieter and more unsettling story: the story of computational complexity. This field is less concerned with what can be solved than with how efficiently it can be solved, and that distinction changes everything. It marks one of the deepest intellectual fault lines of modern thought, one that traces its ancestry to Gödel and Turing but follows a rhythm all its own.

At first glance, complexity theory can look like a dry technical aside, a footnote buried in computer science textbooks. In reality, it cuts straight to the limits of computation. Imagine working on a jigsaw puzzle. If someone hands you the finished picture, checking that it is correct takes almost no effort. But assembling the puzzle from scratch can range from trivial to excruciating. Some puzzles practically solve themselves. Others demand what feels like endless trial and error.

This gap between 'finding' a solution and 'checking' one lies at the heart of the most famous open problem in computer science: P versus NP. Here, the class P consists of problems that a computer can solve efficiently, meaning in polynomial time. NP consists of problems for which a proposed solution can be efficiently verified, even if finding that solution may be difficult. The central question is deceptively simple: if a solution can be checked quickly, must it also be possible to find it quickly? Most researchers suspect the answer is no. If P is not equal to NP, then there exist vast families of problems that resist efficient solution even though their answers, once discovered, are easy to confirm. These are not mathematical curiosities tucked away in theory. They include problems at the core of logistics, drug design, machine learning, and cryptography. Complexity theory, in this sense, draws a sharp boundary between what is practically achievable and what remains computationally out of reach.

What makes this boundary especially fascinating is how abruptly it can appear. Many computational problems exhibit behaviour reminiscent of phase transitions in physics. Below a certain threshold, solutions are plentiful and easy to find. Cross that threshold, and the same problem becomes suddenly intractable. A small change in parameters can flip a system from order to chaos. This suggests that complexity is not just about processing speed or hardware muscle, but about the deep internal structure of problems themselves.

One of the quieter architects of this insight was Leonid Levin, who independently of Stephen Cook developed the theory of NP-completeness. NP-complete problems are the most difficult members of NP in the sense that any other NP problem can be efficiently transformed into them. Solve one efficiently, and you solve them all. Levin's contributions, long underappreciated outside the Soviet sphere, are a reminder that foundational ideas often transcend political and cultural boundaries, even when recognition does not.

The implications of complexity theory extend across science, mathematics, and everyday life. Modern cryptography depends on the assumed hardness of certain problems. The security of digital communication rests on the belief that cracking a code is not merely difficult but infeasible within any realistic timeframe. This assumption has held for decades, but it is now under pressure.

And the reason being that quantum computing that threatens to redraw the complexity landscape.

In 1994, Peter Shor showed that a sufficiently powerful quantum computer could factor large numbers exponentially faster than any known classical algorithm. Since much of today's cryptography relies on the difficulty of factoring, Shor's algorithm exposed a potential fault line in the foundations of digital security. As a result, researchers are racing to develop quantum-resistant cryptographic systems, turning complexity theory into a live arena where abstract limits meet urgent technological stakes.

Seen this way, computational complexity is not a minor technical concern. It is a map of the possible and the impossible, a guide to where intelligence, whether human or artificial, must slow down or stop altogether. Like Gödel's incompleteness results before it, complexity theory reminds us that limits are not failures. They are features of the landscape and understanding them is the first step towards navigating what lies beyond.

On a philosophical level, complexity theory adds an unexpected turn to the Gödel–Turing story. Gödel showed that some truths are forever unprovable within a formal system. Turing showed that some problems are undecidable, no matter how ingenious the algorithm. Complexity theory goes a step further. It tells us that even when a problem can be solved in principle, the time required may be so vast that, for all practical purposes, the problem is unsolvable. This is the difference between impossibility in theory and impossibility in practice, and it reshapes how we think not only about machines but about nature itself. For example, protein folding, weather systems, and economic markets all exhibit forms of computational complexity that resist reliable prediction. The issue is not a lack of equations or data, but the sheer explosion of possibilities that must be navigated. Complexity dynamics, in this sense, mark the edge of what can be known and force us to rethink what it really means to 'solve' a problem, not in abstraction, but under the hard constraints of time and resources.

This aspect also raises a concerning possibility. What if the universe of mathematical truths, those shimmering Platonic ideals we imagine as existing 'out there,' is forever beyond our reach? Not because these truths are false or inconsistent, but because uncovering them would require more computational power than the universe itself can supply. This is not mere speculation. Complexity theorists such as Scott Aaronson have argued that if P ≠ NP, as most believe, the consequences for mathematical Platonism may be even more severe than Gödel's incompleteness theorems.

Picture mathematics as a vast archipelago, with islands of elegant theorems shining in the distance. The islands exist, and their beauty is real, but the seas between them are computationally treacherous. No matter how fast your ship, the waters may be too deep and too vast to cross. Human understanding becomes heroic and tragic at once, always reaching, always constrained, always incomplete.

Yet this story does not end in abstract mathematics. It curves inward, towards the nature of consciousness itself. Douglas Hofstadter, whose work is inseparable from explorations of self-reference, pursued this connection most famously in *Gödel, Escher, Bach* and *I Am a Strange Loop*. His central claim is audacious: consciousness, the elusive sense of 'I,' is itself a strange loop.

Hofstadter defines a strange loop as a situation in which, by moving through the levels of a hierarchical system, one unexpectedly arrives back where one began. In the brain, he suggests, patterns of activity model not only the external world but also themselves. Mental processes reflect on mental processes, forming recursive layers of self-representation. In this looping mirror, a stable sense of self emerges.

To imagine this more vividly, consider the mind itself as a recursive playground, where reflections interact and feed back into one another. And in this hall of mirrors, each reflection contains reflections of its own. Like Gödel's self-referential sentences, which speak about their own provability, the brain builds models that refer back to the model-builder. Out of this tangled hierarchy arises self-awareness. From this perspective, the self is not a fixed object, but a dynamic pattern, a process sustained by recursion and feedback. It is resilient and fragile at the same time, more verb than noun, more dance than thing.

Moreover, this idea is no longer confined to philosophical speculation. Cognitive science increasingly points to the importance of recursive representations, brains that build maps of their own maps, as a foundation for complex thought and awareness. Strikingly, these modern theories echo ancient insights from traditions such as *Advaita Vedanta* and *Dzogchen*, which long described the self as a mirage, a construct arising from the mind mistaking its own reflections for something solid. Here, contemporary science and ancient philosophy meet in a strange loop of their own.

So, from Gödel's theorems in formal logic to the lived mystery of consciousness, self-reference runs like a hidden thread. These loops are not curiosities or glitches. They may be the deep grammar of mind and meaning, the mechanism by which matter gives rise to experience and systems encounter their own limits.

Gödel's legacy, then, is far more than a technical result in mathematical logic. It is an invitation to confront the boundaries of knowledge itself. What can be known? What can be computed? And perhaps most unsettling of all, how does a system come to know itself? The answers, if there are any, seem forever entangled in loops that reflect, recur, and beckon us deeper into the strange and unfinished architecture of understanding.

The Mathematical Multiverse

Imagine, for a moment, that mathematics is a bustling city of spires, domes, alleyways, and secret underground tunnels. This is more than a metaphor; it is a manifesto. Mathematician Joel David Hamkins champions what he calls the 'mathematical multiverse.' Think Marvel, but with axioms instead of Avengers. Seen this way, mathematics is not the discovery of one universal truth but the construction of many consistent, and sometimes contradictory, worlds. In some of these worlds, the axiom of choice holds. In others, it does not. In one, the continuum hypothesis is true; in another, it fails. Yet each world is internally coherent.

Even Gödel, whose name is inseparable from incompleteness, leaned towards this pluralistic vision. In his later years, he became fascinated by large cardinals, infinities so vast that they tower over the merely countable like Himalayan peaks over foothills. These exotic axioms, though unprovable from standard set theory, may settle long-standing undecidable questions. Gödel suspected they might one day illuminate the very problems his theorems had declared unreachable.

Additionally, the roots of this pluralism reach back to the mid-twentieth century, when Saunders Mac Lane, co-founder of category theory, remarked that mathematics was not a temple but a metropolis. There is no single sacred blueprint, only neighbourhoods, crosswalks, and intellectual subway lines. Some districts even allow algebra and geometry to whisper to one another across abstract bridges of analogy. Mac Lane urged mathematicians to consider a different kind of foundation, one that focused not on sets and their elements but on functions and relationships. As he put it:

> 'The standard "foundation" for mathematics starts with sets and their elements. It is possible to start differently, by

> axiomatising not elements of sets but functions between sets. This can be done by using the language of categories and universal constructions.'

Category theory became a kind of linguistic Rosetta Stone for modern mathematics. Where traditional approaches fixate on proving individual theorems, category theory steps back and asks how entire structures relate. It is less like cataloguing birds and more like understanding the shape and motion of a flock. In that sense, it functions as a grammar that allows different mathematical dialects to speak to one another.

Mathematicians such as William Lawvere pushed this relational view as a challenge to classical foundational thinking, proposing that mathematics is fundamentally about structure rather than absolute substance. In this sense, category theory becomes a quiet admission that truth is contextual, partial, and perspective-dependent. Even philosophers such as Willard Van Orman Quine treated mathematics and science not as access points to eternal Platonic realms but as parts of a single, interconnected 'web of belief,' grounded in usefulness and coherence.

But this abstract picture raises a deeper question: if mathematics is woven into our web of beliefs, how does it show up in ordinary experience?

Now picture yourself at a café in Paris, watching pigeons trace elegant aerial patterns. As you sip your coffee, you notice the symmetry of their motion, and your mind, without conscious effort, performs a rough trigonometric calculation. Pause here. Are you merely applying a clever mental trick, or have you brushed against something fundamentally real? This is where the Quine–Putnam indispensability argument enters, with the inevitability of a well-formed proof.

Quine famously argued that 'to be is to be the value of a bound variable.' The slogan hides a serious claim. If our best scientific theories quantify over numbers, sets, or functions, then we are committed to their existence. Quine believed our metaphysical commitments should match the entities required by our most successful theories. If physics relies on Hilbert spaces or Lie algebras, then those structures have earned a seat at the ontological table.

Hilary Putnam joined this line of thought with his 'no miracles' argument. If science works as well as it does, then the mathematics that underwrites it cannot be mere ornamentation. Its effectiveness would otherwise be inexplicable. As Putnam suggested, the success of science would be miraculous if it were not at least approximately true. And if mathematics is woven into the very fabric of scientific explanation, denying its reality is like admiring a Renaissance painting while insisting the canvas is an illusion.

The main takeaway is that mathematics is not just about numbers, nor is it limited to describing planetary orbits or neural networks. It is the language of structure. From gauge symmetries in quantum field theory to Hilbert spaces in quantum mechanics, the indispensable frameworks of modern physics are not primarily numerical; they are algebraic. Try to remove them, as Hartry Field attempted in *Science Without Numbers*, and the result is a theory so unwieldy it makes a Victorian steam engine look streamlined.

But the story runs even deeper. Much of modern science relies on second-order logic, which quantifies over properties and relations, not just individual objects. To interpret such logic, one needs sets, collections that exist at an abstract level but exert real explanatory force. If you are discussing the logic of science at all, you are already immersed in a form of set-theoretic realism, whether you acknowledge it or not.

This is where category theory enters as a new contender for foundational status. Some philosophers suggest that it might replace set theory with something even more abstract, yet conceptually leaner. If that shift succeeds, realism would move away from objects like numbers and towards structures such as morphisms and functors. But this structural realism does not free us from ontological commitment; It merely changes what we are committed to.

Then comes another twist: mathematics not only describes the universe; it shapes how we perceive it. We never encounter raw cosmic chaos. We encounter coordinates, vectors, and probability distributions. Therefore, mathematics is not simply a lens placed before reality; it functions more like the retina through which reality becomes visible at all.

As G. K. Chesterton once put it:

> 'The difference between the poet and the mathematician is that the poet tries to get his head into the heavens while the mathematician tries to get the heavens into his head.'

This brings us back to the Quine–Putnam argument which calls for intellectual honesty. If our best theories rely on mathematics not merely for elegance or convenience but for genuine explanation, then denying the existence of mathematics is like insisting that Shakespeare's plays do not require language. Some will still argue that mathematics is a useful fiction, a sophisticated metaphor we deploy to navigate the world. Perhaps so. But even fictions have authors. And, as Quine and Putnam suggested, the author may well be the universe itself, expressing its structure in symbols that only mathematics can articulate.

Solomon Feferman, a leading logician and philosopher, pushed this idea further with what he called 'conceptual structuralism.' Mathematics, on this view, is a human construction, but not an arbitrary one. It is disciplined by internal logic, yet open to unbounded creativity. This shift, from oracle to artisan, reshapes the role of the mathematician. No longer a scribe copying divine truth, the mathematician becomes an artist, engineer, and inventor at once.

Coming full circle, Gödel's incompleteness theorem does not show that mathematics is broken. It shows that it is alive. Had David Hilbert's dream of a complete and decidable system succeeded, mathematics might have ossified into bureaucratic routine. Instead, Gödel left the gates open. Building on this openness, Gregory Chaitin developed algorithmic information theory and showed that randomness appears at the very foundations of mathematics. This implied that some truths are true for no deeper reason. They are irreducible and no shorter explanation exists. Like snowflakes, they resist compression. As Chaitin remarked, mathematics is closer to biology than to logic. On the other hand, we also have theorems whose shortest proofs would require more computational resources than the observable universe could ever supply. These truths exist, yet no feasible reality could verify them.

Strikingly, this theme of irreducible limits appears across disciplines. In physics, Heisenberg's uncertainty principle tells us that even nature withholds complete knowledge. In cosmology, black holes hide their interiors behind event horizons. In computer science, Turing proved that no machine can solve every problem. In meteorology and ecology, chaos theory shows how simple systems can behave unpredictably.

Another turning point of the twentieth century came when Paul Cohen proved that both the continuum hypothesis and the axiom of choice are independent of the standard Zermelo–Fraenkel axioms. In other words, one can construct fully coherent mathematical universes in which these propositions are true, false, or neither provable nor disprovable. These results have concrete consequences, shaping how functions behave, how sets are constructed, and how infinity itself is understood.

At the same time, far from the mainstream spotlight, mathematicians such as Alexander Grothendieck were quietly reimagining the foundations of mathematics altogether. His development of topos theory made it possible for different logical systems, including intuitionistic and modal logics, to exist inside mathematics itself. One can now study mathematical worlds where classical logic is optional, where negation behaves differently, and where truth is not governed by a single universal rulebook.

Grothendieck embraced this abstraction with characteristic audacity. In a letter to Robert Brown dated 5 May 1982, he remarked:

> 'The introduction of the cipher 0 or the group concept was general nonsense too, and mathematics was more or less stagnating for thousands of years because nobody was around to take such childish steps.'

The phrase 'abstract nonsense,' coined jokingly by Norman Steenrod, refers to the highly general methods of category theory and homological algebra, where proofs rely on universal properties rather than element-by-element constructions. Despite its dismissive sound, the term is used affectionately. Grothendieck took it seriously, arguing that such nonsense was foundational in exactly the same way that zero or group theory once were. What looked gratuitously abstract turned out to be the meta-language capable of unifying vast regions of mathematics.

Yet abstraction does more than unify as it also creates the conditions for radical conceptual change. This creative breaking point is precisely where the Colombian mathematician and philosopher Fernando Zalamea locates major breakthroughs in mathematics, describing them as acts of 'transcendental creativity.' These are moments when new worlds emerge from the rupture of old frameworks. Mathematics, in his view, advances not only through deduction but through visionary leaps. For instance, Bernhard Riemann's introduction of manifolds in the nineteenth century was not merely a technical extension of geometry. It opened the conceptual path to curved spaces, later essential to Einstein's theory of general relativity. Georg Cantor's discovery that some infinities are larger than others shattered centuries of assumptions and ignited philosophical debates that still rage today.

Zalamea also draws attention to less celebrated transitions, such as the move from discrete symmetry groups to the continuous symmetries of Lie groups, which now sit at the heart of quantum mechanics and particle physics. Each such shift marks a moment when mathematics breaks free from its own constraints and invents new ways of seeing. These are not refinements at the margins but transformations at the core, moments when imagination and rigour collide to produce something genuinely new.

From all these observations, we can conclude that mathematics is not a finished edifice advancing towards a final answer, but a living practice shaped by rupture, imagination, and uncertainty. It does not shy away from paradox; it draws strength from it. Far from being a single, settled universe, mathematics unfolds as many coexisting worlds—some rigorously structured, others playfully speculative—each stretching the limits of what can be known. Moreover, it asks questions it knows cannot be fully resolved and thrives precisely in that tension. And in doing so, mathematics becomes more than a subject of study and instead stands as an ongoing invitation to think beyond certainty and to remain open to realities yet to be conceived.

> 'Nothing in biology makes sense except in the light of evolution.'
>
> — **Theodosius Dobzhansky**

chapter 5
bio absurdity

In 1973, Lynn Margulis proposed a theory so radical that fifteen scientific journals rejected it outright. Within a few decades, that same idea would rewrite biology's rulebook. Her claim was startling in its simplicity: mitochondria, the tiny power plants inside nearly every cell in your body, were not originally part of those cells at all. They began as free-living bacteria that were engulfed by early single-celled organisms. Instead of being digested, they stayed, forming a partnership so intimate that it transformed the meaning of life itself.

This was Margulis's endosymbiotic theory, and it was initially dismissed as heresy by a scientific culture devoted to clean categories and rigid hierarchies. Today, it is foundational biology. We now know that mitochondria and chloroplasts in plants carry their own DNA, clear evidence of an independent evolutionary past. Genetic analysis traces their lineage to alphaproteobacteria, relatives of the Rickettsiales, a group that includes the bacteria responsible for typhus. Thus, what was once an invader became family, and with that realisation, the tidy boundary between 'self' and 'other' begins to dissolve.

This story of microbial merging reflects a broader truth about life's refusal to fit into human-made boxes. What, after all, is a species? Biologists still debate more than two dozen definitions, from the biological species concept, which focuses on reproductive compatibility, to the phylogenetic species concept, which emphasises shared ancestry. Each comes with exceptions that undermine its authority. Coral reefs can stretch for kilometres while being composed of genetically identical polyps functioning as a single organism. Lichens, intimate unions of fungi and algae, challenge the very idea of what an organism is. Life is fluid, layered, and persistently ambiguous.

Long before microscopes revealed the microbial world, many cultures sensed this fuzziness intuitively. The Māori of Aotearoa New Zealand view humans, *tangata*, not as masters of nature but as inseparable from the environment, *taiao*, woven into a shared living fabric. Jain philosophy in India speaks of *nigoda*, innumerable microscopic life forms permeating existence. In fact, similar perspectives appear across cultures, which understand life not as a collection of isolated individuals, but as a web of interdependence, constant interaction, and transformation. This way of viewing the universe imagined it as teeming with invisible life long before modern biology confirmed such ideas.

One of the clearest illustrations of this interdependence can be seen in the relationship between the bobtail squid and its luminous bacterial partners, *Vibrio fischeri*. The squid shelters the bacteria in a specialised pouch and supplies them with nutrients. In return, the bacteria produce bioluminescence that masks the squid's shadow in moonlit waters, protecting it from predators below. Here, cooperation and survival are not opposites but intertwined aspects of the same evolutionary strategy.

This kind of improvisation reveals nature's true working style. The French biologist François Jacob described evolution as a *bricoleur*, that is, a tinkerer assembling workable solutions from whatever materials happen to be available, rather than an engineer executing a flawless design. The narwhal's famous tusk, once assumed to be a weapon, is actually a highly sensitive sensory organ packed with millions of nerve endings. The giraffe's recurrent laryngeal nerve takes an inexplicable two-meter detour around the heart, a historical accident inherited from fish ancestors that never had necks to begin with. All these examples show that there are no absolute boundaries, no perfectly sealed categories, only ongoing negotiation. In other words, life does not write pristine equations; it composes sprawling, unfinished genealogies. This is the deeper lesson of Margulis's mitochondria, glowing squids, and evolutionary detours.

American ecologist Frank Edwin Egler put this sentiment with sharp precision when he said:

> 'Nature is not more complicated than you think; it is more complicated than you can think.'

Ultimately, what connects microbes and mammals, ancient wisdom and modern science, is a shared refusal of simplicity. It urges us to rethink what we mean by self, life, and other, and to adopt a humility that feels increasingly urgent in a world addicted to sharp borders and easy certainty.

So, when life seems chaotic or ill-defined, remember that its power lies precisely there. This is not a story of disorder but of connection. Not a failure of design but a triumph of adaptation. In embracing life's beautiful ambiguity, we may yet learn how to live more wisely together.

Nature Refuses Neat Boxes

In the emerald shade of the Western Ghats, the honey-hunters of the Kattunayakan tribe speak of bees that are brothers but not brethren. Different kinds of wild *Apis* forage on the same blossoms and build hives in the same trees, yet they never mix, never interbreed. It is a poetic insight from the forest floor, and it happens to strike at one of biology's deepest puzzles: what exactly is a species?

On the first look, the answer seems obvious. Species appear as self-evident categories, butterflies pinned in neat rows, lions in one box, and tigers in another. Nature looks orderly and well sorted. However, even Charles Darwin found this apparent clarity deceptive. Writing to his close friend Joseph Hooker in December 1856, he confessed his exasperation:

> 'It is really laughable to see what different ideas are prominent in various naturalists' minds, when they speak of "species" in some resemblance is everything and descent of little weight-in some resemblance seems to go for nothing. Creation the reigning idea-in some descent the key-in some sterility an unfailing test, with others not worth a farthing. It all comes, I believe, from trying to define the undefinable.'

That single word, undefinable, reveals Darwin's refusal to treat species as fixed, absolute entities. For him, species were not timeless essences but temporary waypoints in an ongoing evolutionary process. This position became even clearer as he concluded *On the Origin of Species*, where he argued that much of the confusion surrounding classification arose from the mistaken search for an inner essence. As he wrote:

> 'Systematists will be able to pursue their labours as at present, but they will not be incessantly haunted by the shadowy

> doubt whether this or that form be in essence a species. Hereafter, we shall be compelled to acknowledge that the only distinction between species and well-marked varieties is that the latter are known, or believed, to be connected at the present day by intermediate gradations, whereas species were formerly thus connected. Hence, without quite rejecting the consideration of the present existence of intermediate gradations between any two forms, we shall be led to weigh more carefully and to value higher the actual amount of difference between them . . . we shall at least be freed from the vain search for the undiscovered and undiscoverable essence of the term species.'

A century later, Ernst Mayr gave this dilemma a name: the species problem. Today, biologists juggle more than two dozen competing species concepts, each trying to trace the wild contours of life, and each inevitably spilling beyond its own boundaries.

The most familiar is the biological species concept. If two organisms can produce fertile offspring, they belong to the same species; if they cannot, they do not. The rule sounds clean and decisive, until you ask the bacteria. They reproduce by fission, not courtship, and cheerfully ignore the entire premise.

Next comes the phylogenetic species concept, which treats DNA as a genealogical record. Here, species are defined not by who can mate with whom but by shared ancestry. A species becomes the smallest branch on the evolutionary tree that carries a distinct genetic signature.

Then there is the ecological species concept, which defines a species by the role it plays in its environment. Two organisms may look nearly identical, but if one inhabits a cactus and the other a coral reef, evolution may regard them as different actors in nature's drama.

There is also the older and more intuitive morphological species concept, which classifies species based on visible traits such as feathers, fins, or fangs. Thus, if it looks like a duck and quacks like a duck, it probably is one. It is a taxonomy by silhouette.

All these concepts represent only a handful of attempts at classification. Biologists have proposed more than two dozen such definitions, each capturing a different aspect of biological reality. Other prominent approaches include the evolutionary species concept, which treats species as lineages with their own long-term trajectories. The recognition concept focuses on unique mating systems. The genotypic cluster concept identifies species as genetic groupings separated by gaps. Additionally, some scientists adopt an explicitly pluralistic approach, accepting that no single definition can capture nature's endless improvisations.

Yet none of these concepts can claim the final word. The British philosopher John Dupré suggests that the problem lies not in our models but in reality itself. Perhaps nature resists tidy classification because it is fundamentally plural. Life branches, loops, and recombines without regard for our categories.

Consider the Ensatina salamanders of California. Their populations form a horseshoe around the Central Valley. Neighbouring groups interbreed freely, like dialects blending across villages. But at the southern end of the ring, where the two arms meet, the populations can no longer produce viable offspring. So, are they one species or two? This question dissolves under its own weight.

Let's look at another example. Greenish warblers migrating around the Himalayas exhibit gradual changes in song from one region to the next, with populations at opposite ends eventually unable to recognise each other's calls. These cases are not exceptions; they are lessons. Thus, speciation is not a sharp boundary between two colours. It is a gradient, where each shade blends into the next until the ends no longer resemble each other.

In this context, the most radical perspective comes from microbiology. Horizontal gene transfer, in which genes move sideways between organisms like gossip at a village well, shatters the classic image of the Tree of Life. Bacteria not only inherit genes from their ancestors; they trade them with strangers. This is how they evolve antibiotic resistance and metabolise new substances. The biologist W. Ford Doolittle has even suggested that prokaryotic species may not exist in any meaningful Darwinian sense. This shows that microbial evolution is better understood not as a single tree but as a banyan grove, where tangled branches and aerial roots intertwine over time.

This logic of entanglement does not stop at the microbial fringe; it reaches into the very core of our own cells. Nowhere is this more vividly embodied than in mitochondria, the energy-producing structures inside our cells. Once free-living bacteria, they entered into a symbiotic partnership with ancestral cells around two billion years ago. It was not a conquest but a merger, a cellular alliance whose descendants power every breath we take. Their small, stubborn genomes still carry echoes of their former independence. Therefore, even humans are mosaics of this history as nearly twenty per cent of our genome carries the traces of ancient viral invasions, written into our DNA like palimpsests.

So what, then, is a species? Perhaps the better question is how species come into being. Sometimes the process is dramatic, as when a mountain range splits a population in two. Other times it is subtle, as when corals spawn on different nights or flies respond to different rhythms. Over time, these separations accumulate, and a new lineage emerges. Taken together, a species is not a Platonic ideal floating in a realm of perfect forms. It is a historically contingent, biologically fuzzy, and culturally convenient label. A name we give to a lineage at a particular moment, like freezing a dancer mid-spin. And perhaps that is the point: life does not draw lines, but we do.

When Organisms Defy Singularity

In the mist-shrouded forests of Meghalaya, the Khasi people use living root bridges to create pathways across rivers and gorges. These bridges are neither built in the conventional sense nor simply grown like trees. Over decades, and sometimes centuries, villagers patiently guide the aerial roots of the rubber fig (*Ficus elastica*) across rivers, weaving them with moss, stones, and the quiet labour of microbial communities. The result is a structure that grows stronger with time. Moreover, no single person can claim authorship of these bridges as they emerge from collaboration between human intention and living systems.

This quiet innovation offers a striking lesson: individuality in nature is not a fixed fact but something shaped by context, time, and scale. The same insight now echoes through biology, where once-stable categories such as species and organism are beginning to fray.

Take lichens, long mistaken for simple plants clinging to rock or bark. In the late nineteenth century, the Swiss botanist Simon Schwendener proposed a radical idea: lichens are symbiotic partnerships, typically between a fungus and an alga or cyanobacterium. The fungus builds the protective structure and extracts minerals from the substrate, and the photosynthetic partner converts sunlight into nourishment. Neither survives alone, yet together they form a new and durable way of life.

Modern genetics has deepened this picture. Many lichens turn out to be multispecies collectives, hosting multiple fungi, bacteria, and even yeasts. Some fungal partners have become so dependent on their companions that they cannot reproduce sexually without them. In certain cases, fungal genomes contain genes likely acquired through horizontal transfer from bacteria, a rare crossing of boundaries between domains of life. What, then, is a lichen? Viewed in this light, it is neither a single organism nor a simple community seems quite right.

A similar ambiguity appears in the slime mould *Dictyostelium discoideum*. For most of its existence, it lives as a solitary amoeba, roaming independently and feeding on bacteria. When food runs out, thousands of these cells aggregate into a slug-like body that behaves as a multicellular organism. Some cells sacrifice themselves to form a stalk, lifting spores that disperse and begin the cycle again. These slugs may contain genetically unrelated individuals, and some cells even 'cheat' by avoiding the sacrificial role. Yet the collective endures. So, is *Dictyostelium* one being or many? The answer depends on how and when you look, much like ancient Indian reflections on the shifting nature of the self.

Even more intimate is the ecosystem each of us carries within. The human body hosts roughly thirty-eight trillion bacterial cells, a number comparable to our own, along with vast populations of viruses, fungi, and archaea. These organisms are not passive passengers. They shape digestion, tune immune responses, influence mood, and affect brain chemistry through the gut-brain axis. Growing evidence suggests that the microbiome can sway emotions and decisions, implying that the 'self' may be less singular than we once believed.

The same pattern appears in forests. Research has shown that trees exchange nutrients and chemical signals through underground mycorrhizal networks, often called the 'wood wide web.' Fungal threads connect the roots of different trees, allowing them to warn one another of pests or share resources in times of stress. A tree, then, is not an isolated organism but part of a larger, cooperative system. Seen this way, the forest itself begins to resemble a superorganism.

Here, an important question naturally follows: What is an individual? Like the living root bridges of Meghalaya, the composite body of a lichen, or the shifting unity of a slime mould, life resists singular definitions and refuses to offer clear-cut answers. It unfolds as a process of cooperation, conflict, and continual transformation. Put simply, individuality is inseparable from symbiosis.

Recognising this, biologist Scott Gilbert and his collaborators have argued that the traditional image of the autonomous organism no longer suffices. In its place, they propose the concept of the holobiont: a composite being made up of a host and its many microbial partners. Together, this ensemble functions as a single biological unit, even though it is woven from multiple lineages. From this perspective, biology looks less like a collection of individuals and more like a jam session, collaborative, contingent, and alive to improvisation.

Once we adopt this lens, the implications ripple outward, challenging how we see life at every scale. Our bodies, forests, coral reefs, and even lichen-covered stones are not mere assemblies of parts. Instead, they are dynamic patterns of relationship and mutual becoming. This insight quietly undermines a long-standing Western fixation on individuality, shaped by Enlightenment ideals and reinforced by political metaphors of autonomy and self-sovereignty. Under close inspection, the boundaries that once seemed solid begin to dissolve, revealing life as fluid and relational. Thereby, biology's most radical claim may be this: the bounded, self-sufficient individual is an abstraction. What we call life is a continuous process of co-creation, a web of interdependence stretching from microbes to ecosystems.

This realisation does more than humble us. It invites wonder to think that perhaps life's deepest truth is that the self is never truly alone and cannot exist in isolation. We are threads in a vast net of relations, endlessly reflecting, sustaining, and shaping one another.

Adaptation Means Absurdity

Up to this point, we have seen how life resists fixed categories and how uncertainty is not a flaw in biology but one of its defining features. This non-absolute nature of biology becomes clearest when we look closely at adaptation itself. Evolution does not sculpt perfect or optimally designed organisms. It behaves more like an artisan constrained by memory, improvising workable compromises from whatever materials history has left behind and whatever pressures the present demands. This is not divine engineering but a slow dance of tinkering.

In ancient Indian sculptural practice, it is often said that the finest form is not created but revealed, as the sculptor removes what does not belong. Life unfolds in much the same way. Nature chisels, repurposes, and reshapes without foresight, guided instead by constraint and circumstance. One of the clearest expressions of this process is exaptation, a concept introduced by Stephen Jay Gould and Elisabeth Vrba. Unlike adaptation, where a trait evolves in response to a specific pressure, exaptation describes the co-option of existing traits for entirely new purposes. For instance, bird feathers, now inseparable from the idea of flight, likely originated for insulation or display. Fossils such as *Archaeopteryx* show stiff, pennaceous feathers poorly suited for powered flight but well suited for warmth or visual signalling in the dappled light of Mesozoic forests. Flight, in this view, was not the original goal but an evolutionary afterthought, a fortunate accident. Such repurposing is not exceptional; it is fundamental to life's history. The malleus and incus of the mammalian middle ear once formed part of the jaw joint in reptilian ancestors. The swim bladder of modern fish descends from primitive lungs that enabled early fish to survive in oxygen-poor waters. Even the crystallin proteins that give our eyes their clarity began as metabolic enzymes before being recruited for vision. Like Renaissance artisans who carved saints from fragments of Roman stone, evolution, as François Jacob observed, works not as an engineer but as a bricoleur, assembling the new from the remnants of the old.

It is also worth noting that evolution's creativity comes with costs. Many genes perform multiple functions, a phenomenon known as pleiotropy. When a gene that benefits one trait harms another, antagonistic pleiotropy arises. George C. Williams proposed that genes promoting strength and fertility in youth may also hasten decline in old age. Calcium strengthens bones early in life but contributes to arterial stiffening later on. It is a poignant trade-off, reminiscent of Kalidasa's insight as a classical poet and dramatist, that the brilliance of youth must eventually yield to the shadow of age.

These compromises extend far beyond human biology. The peacock's extravagant train, celebrated in Sanskrit poetry and Mughal miniatures, dazzles potential mates but hinders flight and increases vulnerability to predators. Fast-growing trees such as teak and sal dominate monsoon forests, yet their speed comes at the cost of softer, more fragile wood. Even the immune system, vigilant by design, can become destructive when its response exceeds its purpose, turning against the body in autoimmune disease.

Perhaps the most revealing example of this is the vertebrate eye, often praised as a masterpiece of design. Its photoreceptors are wired backwards, forcing light to pass through layers of neurons and creating a blind spot. Cephalopods such as octopuses evolved a cleaner solution, with forward-facing photoreceptors and no blind spot at all. The same problem was solved along different paths, one shaped by compromise, the other by a cleaner slate. Biology's story, then, is not one of perfection but of persistence. Evolution does not pursue ideal forms; it settles for what works, for now, within the constraints of what already exists.

This principle becomes even more apparent when we follow the body's wiring rather than its visible structures. If you were asked to design a highway from the brain to the voice box, you would draw a straight line. Evolution did not. Instead, it routed the recurrent laryngeal nerve on an elaborate detour through the chest cavity.

In fish, this nerve ran directly from the brain to the gill, a perfectly sensible arrangement. But as vertebrates moved onto land and necks lengthened, evolution, constrained by embryological history, did not redraw the plan. It simply stretched what was already there.

In giraffes, the result is a striking anatomical absurdity. The nerve descends from the brain into the chest, loops around the aorta, and climbs back up to the larynx, travelling more than fifteen feet to reach a destination only inches away. It is like sending a Mumbai autorickshaw from Churchgate to Colaba via Pune because, long ago, the road happened to pass that way. Biologists call this path dependence, the persistence of historical routes even when better ones are obvious.

Now, consider the human spine. An S-shaped column of stacked bones struggles to keep us upright, even though it was inherited from quadrupedal ancestors who never sat at desks or stood in queues. The result is slipped discs, sciatica, and an entire industry devoted to ergonomic furniture.

These oddities are not flaws in the system; they are the system. The pharynx, for instance, serves both as an airway and a food passage, a risky dual-purpose corridor responsible for millions of choking incidents. Yet this arrangement enables speech. The lowered larynx that lets us articulate also makes it possible for a misplaced samosa to become lethal. Thus, language, it turns out, is a dangerous gift.

Even our ears are archives. The malleus, incus, and stapes, three delicate bones that transmit sound, once formed part of the jaw apparatus of ancient fish. That we hear music today through the transformed remains of a gill arch is a quiet piece of evolutionary poetry.

Nature works in much the same way. It is not a clean-slate engineer but a master of *jugaad*, assembling life through inheritance, accident, and moments of awkward brilliance. Like the gods of the Indian epics, who are shaped by past actions, vows, and battles, nature too operates not with omniscience but with memory. It preserves what works well enough and occasionally stumbles into elegance. Seen this way, imperfection is not a failure but history made visible. Our bodies are not precision-machined devices but living archives, with every nerve and bone bearing traces of what came before. Not always logical. Not always graceful. But unmistakably alive.

When Rules Become Exceptions

From our earlier discussions, it has become evident that biology never signed a peace treaty with predictability. Just when we think we have pinned it down with neat laws and linear graphs, life swerves, spinning into some unexpected harmony, as if to say, 'Nice try.' From the coral reefs of Raja Ampat to the microbial mosaics beneath Antarctic ice, biology resists rigid boundaries.

A classic example is the paradox of the plankton, proposed by ecologist G. Evelyn Hutchinson in 1961. According to the competitive exclusion principle, two species competing for the same limited resources should not coexist for long. One should eventually drive the other out. Yet the open ocean tells a different story. It teems with dozens of phytoplankton species, all drawing from the same basic pool of sunlight, nitrogen, and carbon dioxide.

This suggests that the ocean is not a uniform soup of sameness but a constantly shifting mosaic shaped by seasons, storms, micro-currents, predators, parasites, and chance. And this ceaseless motion injects just enough instability to prevent the system from ever settling, ensuring that no single species can monopolise the stage. Even in relatively calm waters, competition can spiral into oscillations and chaos, allowing many species to persist on a handful of shared resources. Some plankton even partition light itself, using different pigments to capture distinct wavelengths and carving out spectral niches within the same sunlit space.

What emerges, in the end, is not equilibrium but ecological improvisation, layered and perpetually off-script, more like a Raag Darbari alaap that unfolds slowly through shifting moods and modulations, where no single motif dominates forever.

However, this is only part of the story, for biology delights in defying expectations in many other ways. For decades, naturalists leaned on the apparent regularity of the latitudinal diversity gradient, the idea that biodiversity increases steadily from the poles to the equator. The Amazon, with its exuberant abundance, seems to confirm this pattern. But closer inspection complicates the picture. Certain groups of ichneumonid wasps, those elegantly sinister parasitoids, are more diverse in temperate regions than in tropical forests. Beneath Iceland's glaciers, microbial communities thrive with unexpected richness. Some kelp species, celebrated in Japanese literature as *nori no hana*, reach peak diversity not in warm tropical waters but in icy northern seas.

The unified neutral theory of biodiversity (UNTB) pushes this challenge even further. It suggests that species do not always succeed by being better adapted. Sometimes they succeed simply by being present. In other words, survival can hinge on timing, luck, and historical accident. This includes aspects such as: Who arrived first? Who escaped catastrophe? Who happened to disperse in the right direction?

So, for example, in an Amazonian hectare where hundreds of tree species compete for sunlight, the trees that persist do so not because they outperform others, but because they have effectively won an ecological lottery, even though they may have nearly identical birth, death, and dispersal rates as trees that perished. Even coral reefs, those shimmering underwater cities, often show species assemblages shaped as much by randomness as by fierce competition. It is as if nature occasionally rolls dice behind the scenes, letting chance help write the score. Evolution, in this light, resembles a form of democratic chaos. The result is that even the most unremarkable microbe may persist not by dominance or brilliance but by statistical drift.

Then there is evolution itself, the ultimate slow burner, or so Charles Darwin believed. His picture was one of gradual change, unfolding like the steady drift of continents or the quiet passage of generations in a Jane Austen novel. But nature does not always move at that pace. Sometimes, it leaps.

It is precisely at this fault line between gradualism and disruption that the fossil record speaks most clearly. Rather than a continuous flow, it reveals jolts and ruptures. Species appear abruptly, persist for long stretches, and then disappear without ceremony. Palaeontologists Niles Eldredge and Stephen Jay Gould gave this pattern a name: punctuated equilibria. From this standpoint, evolution is not a smooth river but a rhythm, marked by long periods of relative calm punctuated by sudden bursts of change. This resembles *kṣaṇa* in Buddhist philosophy, those fleeting instants of becoming in which reality flashes into form and dissolves again before it can settle.

An additional point to note is that these evolutionary jumps are not driven by divine intervention or exotic outside forces, despite what ancient-astronaut enthusiasts might claim. They emerge from ordinary processes acting under unusual conditions: genetic drift, mutation, and intense selection in small, isolated populations. In fact, it is often the margins, the geographical outliers and ecological misfits, that generate novelty. Horses in North America evolved through a long, largely gradual sequence. Trilobites, by contrast, vanished in a sudden and dramatic crescendo. Molluscs and other shelled organisms frequently show long spans of evolutionary stillness interrupted by brief episodes of rapid transformation. These upheavals tend to occur in remote settings, where small populations face unfamiliar pressures and rare opportunities.

In short, biology does not pledge loyalty to either gradualism or revolution but makes use of both. Like *līlā*, the cosmic play of

Indian metaphysics, evolution unfolds through an interplay of pattern and disruption, of rule and exception.

Evolutionary biology has also drawn attention to ghost lineages, that is, species that leave no direct fossil trace but are inferred from molecular and phylogenetic evidence. Certain early tetrapod groups appear suddenly in Devonian and Carboniferous fossils, yet genetic data suggest their ancestors must have existed tens of millions of years earlier. The lineage leading to modern amphibians is equally striking. Its early history is largely invisible in stone, likely because those organisms lived in environments poorly suited to fossilisation or because their remains have not yet been found. These ghost lineages mark entire chapters of evolutionary history that unfolded unseen.

To this framework, epigenetics adds another layer of complexity. In plants, environmental stress can leave chemical marks that persist for several generations, shaping how populations respond to future challenges without altering their DNA sequence. This acts as a kind of ecological short-term memory, blurring the boundary between inheritance and experience.

From this analysis, we can conclude that biology, and nature at large, is not a ledger of fixed outcomes or a rigid code etched in certainty. It unfolds as a living field of possibility, shaped by improvisation, accident, and chance. Patterns emerge, fade, and reappear in altered forms, not because nature seeks perfection, but because it remains open to variation. And even where order exists, it is primarily provisional and porous, allowing life to thrive not as a contest of winners, but as an unfinished performance in which many forms briefly take the stage.

The Engines of Evolution: Conflict and Cooperation

For a long time, evolution was portrayed as a brutal gladiatorial arena, a relentless contest in which only the strongest or luckiest survived. Tennyson's phrase 'nature red in tooth and claw' captured this mood perfectly: life as a ruthless struggle, with predators, prey, and competition taking centre stage. However, when we look more closely, a far more intricate drama comes into view. Alongside conflict, evolution is filled with unlikely alliances, covert partnerships, and power-sharing arrangements that gradually turned rivals into relatives.

Evolutionary biologists John Maynard Smith and Eörs Szathmáry described these moments as 'major transitions,' episodes in which life fundamentally rewrote its own rules and reorganised itself into more complex forms. DNA molecules that once existed as independent strands clustered together into chromosomes, functioning less like solo performers and more like coordinated ensembles. Ancient bacteria, once entirely self-sufficient, merged to form the first complex cells. From this union emerged mitochondria, the cell's power plants, which still retain traces of their former independence in the form of their own genomes. These were not casual partnerships but long-term commitments, shaped by tension, negotiation, and mutual advantage.

This cooperation did not stop at the cellular level. Single-celled organisms learnt to coordinate and specialise, giving rise to multicellular life. In social insects, evolution pushed cooperation to extremes. Ant and bee colonies evolved in which most individuals surrender reproduction entirely, devoting their lives to the success of a queen. These systems are astonishingly efficient, but they also depend on regulation and governance. Similar principles operate within our bodies. Immune systems evolved to suppress renegade cells, and hormones arose as internal signalling mechanisms to maintain order and coordination.

This interplay of conflict and alliance echoes ideas far older than modern biology. Chanakya, the strategist behind the *Arthashastra*, advised rulers to convert enemies into allies through shared interests and calculated diplomacy. Nature appears to have mastered this strategy long before humans put it into words.

Consider *Wolbachia*, a bacterium that infects more than half of all insect species. In many hosts, it manipulates reproduction by distorting sex ratios or sabotaging sperm, behaving like a microscopic saboteur. Yet in fruit flies, the same organism offers protection against viral infections. In some nematodes, the relationship has become so intimate that neither partner can survive without the other. Therefore, what began as exploitation hardened into dependence. Even bacteriophages, once seen only as microbial predators, have become allies in medicine, deployed in phage therapy to target antibiotic-resistant bacteria. Additionally, parasites such as *Toxoplasma gondii*, infamous for manipulating host behaviour, carry microbial companions of their own. We have also seen how the ancestors of mitochondria, once free-living bacteria, entered host cells as parasites or opportunists and, over time, became indispensable collaborators. Through the exchange of genes and labour, this partnership came to underpin nearly all complex life.

Lynn Margulis captured this insight succinctly when she observed that life did not take over the world by combat but by networking. The image is fitting. In a Japanese Zen garden, stones do not compete and raked sand is not a battlefield. Each element is positioned in relation to the others, contributing to a quiet pattern of interdependence. Difference becomes design rather than dominance. Nature, over deep time, has sculpted similar harmonies. Former enemies became essential partners and once-lethal intrusions were transformed into vital functions.

Yet evolution does not always resolve tension into harmony. Sometimes it accelerates conflict instead. For instance, the Taricha newts of North America produce tetrodotoxin, a neurotoxin potent enough to kill multiple adult humans. In response, populations of *Thamnophis* garter snakes evolved extraordinary resistance. As the newts grew more toxic, the snakes became more resilient. Each adaptation provoked a counter-adaptation in an ongoing cycle of escalation. This is co-evolution in its starkest form, where species do not merely change, but change in direct response to one another.

Evolution, then, is neither purely cooperative nor endlessly violent. It moves through both modes, sometimes weaving partnerships so deep that conflict disappears, and sometimes sharpening rivalries into high-stakes duels. In other words, life advances not by choosing between harmony and struggle but by continually renegotiating the balance between them.

In his 2001 book, *The Botany of Desire: A Plant's-Eye View of the World*, Michael Pollan wrote:

> 'In the wild, a plant and its pests are continually coevolving, in a dance of resistance and conquest that can have no ultimate victor.'

Seen this way, evolution is best understood not as a simple contest between violence and harmony, but as a relational process unfolding within a dynamic web of life. Its trajectories are shaped by continual adjustment and negotiation, as organisms persist not merely by outcompeting others but by learning how to live through them and alongside them. And the deeper we look into the tree of life, the clearer it becomes that today's predator may be tomorrow's partner, and today's selfish gene may be laying the groundwork for a future alliance.

Does Life Follow a Script?

Every age has its sacred mantra. For molecular biology, it was a three-line verse, elegant in its logic and intoxicating in its clarity:

- DNA makes RNA.
- RNA makes protein.
- Proteins make life.

Coined in 1957 by Francis Crick, half of the duo that unravelled the double helix, this formulation became the central dogma of biology. It carried the gravity of a royal decree, or perhaps a Vedic sutra, recited in laboratories across the world. DNA was the sovereign, RNA the messenger, and protein the obedient servant. This also implied that information flowed in a single direction, and to question that flow was to flirt with heresy.

But nature, like poetry, refuses to be confined to tidy stanzas. The first cracks in this molecular scripture did not come from philosophers but from viruses, those liminal tricksters perched at the edge of life and non-life. In the 1970s, researchers studying retroviruses uncovered something deeply unsettling. These viral interlopers were running the script backwards. Using an enzyme called reverse transcriptase, they converted RNA into DNA. It was like discovering a Sanskrit mantra chanted in reverse and still making sense.

This was not a minor exception; it was a fundamental remix. These reverse-engineering viruses were no marginal curiosities. They had left their marks everywhere, embedded in the genomes of animals, plants, and humans. Nearly half of your genome consists of sequences called retrotransposons, genetic stowaways descended from ancient viruses. Far from being useless debris, many of these sequences have been repurposed.

Consider syncytin, a gene vital to the formation of the human placenta, whose origins lie in an ancient viral invasion of our lineage. What once entered uninvited was gradually woven into our biology, becoming essential to the ability of mammals to carry life within life. In this transformation, an old infection was repurposed into a condition for birth itself.

The viral influence does not end there. The immune system, our inner epic battlefield, borrows strategies that echo these retro-elements. The astonishing diversity of antibodies we produce, capable of recognising pathogens we have never encountered, relies on tightly controlled genetic reshuffling. It is the biological equivalent of composing new verses on the fly from an ancient script.

But this creative revisionism does not stop at immunity; it reaches into the very flow of genetic information itself. Even before RNA is translated into protein—the supposed final act of Crick's dogma—the message is already being revised. Through RNA editing, the message can be chemically altered, sometimes beyond easy recognition. In kinetoplastid parasites responsible for sleeping sickness, RNA editing can insert or delete hundreds of letters.

Humans, too, share this editorial impulse. A process known as A-to-I editing allows cells to tweak RNA messages dynamically. This occurs in thousands of genes and affects processes ranging from neurotransmission to immune signalling. The same stretch of DNA can be read one way in the brain and another in the gut.

Then there is alternative splicing, the genome's own remix culture. A single gene can be cut and reassembled in multiple ways to produce a range of proteins, each with distinct functions. The Dscam gene in the fruit fly, which helps neurons wire themselves correctly, can generate more than 38,000 different proteins from one sequence alone. It is as if a single raga could be rendered in tens of thousands of variations, each carrying its own emotional colour.

However, just when biology seemed to have settled into this expanded, conversational view, along came the prions, the true tricksters of molecular biology. First uncovered in the disturbing pathology of mad cow disease and its human counterpart, Creutzfeldt–Jakob disease, prions defied every familiar category. They were neither bacteria nor viruses. They carried no DNA and no RNA. They were protein, and nothing else. Misfolded proteins, to be precise, capable of persuading normally folded proteins to adopt the same corrupted shape. It was as if one dancer slipped, and the entire troupe began copying the stumble, convinced it was part of the choreography.

For decades, biology rested on the belief that information flowed only through the sacred triad of DNA, RNA, and protein. The discovery of prions disrupted this order by revealing that a protein, lacking any genetic script, could nonetheless carry and transmit information. What emerged was a more expansive view in which information can flow from DNA to protein but also from protein back to nucleic acids or sideways into forms of environmental and cellular memory. Though this idea met fierce resistance, nature offered its own rebuttal. Prions were shown not only to spread disease but, in some organisms, to function as carriers of cellular memory, forcing biology to rethink where information can reside. In yeast, prion-like proteins can confer new traits that persist across generations, functioning as a biochemical memory of past stress. In the neurons of *Aplysia*, the sea slug whose simple nervous system has taught us so much about learning, similar proteins appear to stabilise synaptic changes.

Nobel laureate Eric Kandel showed that these prion-like molecules help keep synapses open, allowing experiences to persist as lasting memories. A first kiss, the scent of rain on dry earth, the exact shade of marigold during Diwali may be encoded not in genes, but in the folded state of a protein, held quietly in the synaptic dusk.

All of this has forced biology to rethink its foundations. DNA was no longer a monarch issuing one-way commands. It was part of an ongoing conversation. Additionally, DNA, RNA, proteins, epigenetic marks, cellular environments, and external signals all participate, negotiating outcomes rather than dictating them. As Sergio Pistoi observed in a *Scientific American* article published in February 2020, the genome is less a command centre than a responsive, adaptive system, shaped as much by dialogue as by decree.

'DNA is not a blueprint: it is a recipe coding for thousands of different proteins that interact with each other and with the environment, just like the ingredients of a cake in an oven. Whereas a blueprint is an exact, drawn-to-scale copy of the final product, a recipe is just a loose plot that leaves much more room to uncertainty. Open a packet of cookies: each one was made from the same recipe and baked in the same conditions, but there are no two that are identical. Look closely, and you will spot hundreds of little differences: a burn here, a chocolate chip there, bumps and lumps appearing in distinct places, all because of chaotic interactions between the ingredients and the environment.

Take two identical twins: they share the same DNA, and their embryos developed side-by-side in the same uterus. Yet, they have different tastes, characters and attitudes, and make different choices in life. When you read the DNA of twins, you find a duplicate copy of the same recipe but two distinct personalities. Not what you would call a fixed plan.

We do not inherit specific instructions on how to build a cell or an organ. Our DNA contains a list of biochemical ingredients (the proteins coded in the genes) and the basic rules for their assembly (some proteins are labelled as "master" and can control the activity of others, while others can start a domino-like cascade of events) but the pieces self-organise into biochemical pathways, cells and tissues without reading a manual. The genetic recipe for a cat will not give an elephant, but you can't read the DNA of an individual and see a Mini-Me of his features.

The long-standing blueprint analogy, with its attached determinism, is a toxic meme we have to fight in the era of genetic consumerism. As long as people believe that our identity and fate are programmed into their DNA, there will be a market for questionable genetic tests aimed at predicting intelligence, music, reading and math abilities, and even sexual preferences.'

This upheaval was followed by a series of equally disruptive discoveries. Inteins were found to excise themselves from proteins and, in some cases, to feed information back into DNA, subtly reversing the presumed direction of genetic flow. Nonribosomal peptides bypass the ribosome altogether, assembling complex molecules through enzymatic choreography rather than direct genetic instruction. Taken together, these findings reveal a genome that does not merely execute a fixed script but improvises, continually riffing through variation and constraint.

This shift in biology reflects a broader change in how we understand the world itself, away from isolated parts and towards systems, networks, and context. Life no longer resembles a one-way molecular highway but a dense roundabout of conversation, where signals loop back, detour, and occasionally break the rules altogether. And as for the central dogma, it was not wrong so much as provisional, a tool that worked elegantly within limits before complexity began to speak over it. Like Newtonian mechanics in a relativistic universe, it still holds locally, even as it fails to govern the whole. The genome, in this light, reads less like a fixed transcript and more like an ancient manuscript under constant revision, its meanings layered, contingent, and responsive to circumstance. What DNA seems to declare at first glance often conceals subtler truths beneath the surface. Errors become memories, invasions become inheritance, and certainty gives way to interpretation. In conclusion, biology has not abandoned order, but it has replaced decree with dialogue, turning dogma into an ongoing dance.

When Determinism Meets Plasticity

From the moment humans first turned their gaze inward, the question of how a single microscopic cell becomes a living, breathing organism has inspired wonder, speculation, and story. Poets, philosophers, and scientists have all tried to name this mystery, each offering a different vision of life's unfolding.

In ancient Indian cosmology, Hiranyagarbha, the golden womb or cosmic egg, symbolised the origin of all creation: a single seed holding the universe in potential. Centuries later, Aristotle proposed the idea of the homunculus, a fully formed miniature human hidden within sperm, waiting to grow. Though separated by time and culture, these ideas circle the same question. Is development the gradual revelation of something already complete or the execution of a precise, predetermined plan?

Early modern biology leaned firmly towards the latter. The journey from fertilised egg, or zygote, to organism was imagined as a carefully choreographed process. Cells differentiated step by step into tissues, tissues assembled into organs, and organs built the whole body. Once a cell chose its fate, that choice was thought to be irreversible. Identity was fixed and the genome ruled as a sovereign, issuing commands that could not be undone. Then, in 1996, this certainty was quietly overturned by a sheep named Dolly.

Dolly was the first mammal cloned from an adult somatic cell, taken from mammary tissue. Scientists transferred the nucleus of this fully specialised cell into an egg whose own nucleus had been removed. Against expectation, the egg did not falter. It developed into a healthy sheep, genetically identical to the adult donor.

The implications of this experiment were profound. Dolly demonstrated that even a fully differentiated cell still contains the complete genetic blueprint for an entire organism. Thus, development was no longer a one-way street from stem cell to specialisation. Cells had not discarded genetic information; they had merely silenced parts of it. This also meant that gene expression could be tuned up or down, much like adjusting a radio to different frequencies.

A decade later, the plot deepened. In 2006, Japanese scientist Shinya Yamanaka showed that adult skin cells could be reprogrammed into pluripotent stem cells using just four proteins, now known as the Yamanaka factors. These induced pluripotent stem cells could give rise to neurons, heart cells, and even reproductive cells, making it seem as though time itself had reversed inside the petri dish. Development, once imagined as a straight march forward, now revealed itself as a flexible path that could bend, rewind, and begin again.

Moreover, this plasticity is not confined to laboratories. In nature, it plays out in striking ways. Consider *Daphnia*, the water flea found in ponds around the world. In predator-free environments, it develops a simple, unremarkable form. But when chemical cues from predators appear in the water, the embryo responds by growing helmets and spines for protection. No genetic mutation is required. The change is driven entirely by environmental information. Put differently, the water flea does not wait for evolution to catch up; it reshapes itself in real time.

Plants offer similar lessons. *Arabidopsis thaliana*, the workhorse of plant biology, can grow in dramatically different ways depending on light, water, and nutrients. Some desert plants can even distinguish between being grazed by goats and being attacked by insects, deploying different chemical defences for each threat. It is as though the plant asks a simple question, 'Who is attacking me today?' and responds with remarkable specificity.

Together, these examples point to a quiet revolution in biology, one that replaces genetic command with interaction and context. Genes do not issue fixed orders but offer a range of possibilities that are shaped and refined through environmental signals, cellular states, and lived history. The old divide between nature and nurture dissolves into a continuous exchange, in which development unfolds through listening, interpretation, and response. Life, in this view, is not the execution of a rigid blueprint but a context-sensitive process, emerging from the ongoing negotiation between genes and the worlds they inhabit.

This view lies at the heart of developmental systems theory (DST), a framework that rejects simple hierarchies in biology. Instead of genes ruling from the top and organisms obeying from below, DST describes development as a network of interactions. Meaning, organisms do not merely execute genetic instructions; they actively shape, and are shaped by, their surroundings. Every moment of development becomes a negotiation, a continuous conversation between organism and environment. And here, epigenetics provides a molecular window into this process. Chemical modifications such as DNA methylation and histone changes act as switches that regulate gene activity in response to experience. A mother's diet, the stress of migration, or the rhythm of seasonal change can leave lasting molecular marks. Strikingly, some of these epigenetic signatures persist across generations, allowing experience to echo down family lines in ways Darwin himself could not have anticipated.

The familiar story of development as a linear, deterministic pathway has given way to a more fluid understanding. Cells and organisms are not machines running a fixed program. They are responsive systems, improvising as they go. Even the embryo is not a passive unfolding but a miniature ecosystem, shaped by chance, history, context, and genetic potential all at once.

This perspective is captured clearly in the 2001 report *New Horizons in Health: An Integrative Approach* by the US National Research Council:

> 'Neither genes nor environment dominates development; rather, there is continual interaction between genes and the environment. Phenotype emerges as a function of this constant dialogue, and any effort to ascribe percentage values to isolated variables is likely to be biologically

meaningless . . . If a genetically determined trajectory is not advantageous, then the ability to adjust in response to a new environmental signal would have an adaptive value.'

In summary, molecules, cells, and environmental cues do not simply follow orders; they co-emerge, influencing one another across time and scale. The unfolding of life is shaped by feedback loops, chance events, and context-sensitive adjustments, producing patterns that are flexible, resilient, and often surprising. Viewed from this perspective, the growth of an organism becomes less a predetermined march and more an evolving story, one that invites curiosity, respect, and wonder at the intricate intelligence embedded in living systems.

Biology's Glorious Messiness

In biology, messiness is not a flaw but is the point. Where physics, in the Dirac tradition, seeks beauty through symmetry and elegance, and chemistry finds reassurance in stoichiometric balance, biology thrives in disorder. And the deeper we look into genes, cells, and ecosystems, the clearer it becomes that life is not governed by a single logic but by many overlapping improvisations.

Over the past decade, biology has begun to question even its most basic assumptions, starting with the idea of individuality itself. For centuries, the organism was imagined as a distinct, self-contained unit, genetically coherent and evolutionarily autonomous. That picture no longer holds. We now understand that you are not a solitary being but a living consortium, a host to trillions of microbes—bacteria, viruses, archaea, and fungi—that rival or even exceed the number of human cells in your body. This microbial community is so fundamental that scientists increasingly describe the human body as an ecosystem wrapped in skin. These companions shape metabolism, immune responses, mood, and even aspects of cognition.

At the cellular level, long-standing ideas about biological boundaries have also fractured. Horizontal gene transfer, especially among microbes, allows DNA to move directly between organisms and has emerged as a major evolutionary force. Genes no longer pass neatly from parent to offspring but can jump across species, turning the tree of life into something closer to a tangled rhizome. Even the human genome bears traces of ancient viral invasions, some of which have been repurposed for essential functions such as placental development.

CRISPR has further transformed how we think about life. Known today as a powerful gene-editing tool, it originated as a bacterial defence system against viral attack. Its discovery revealed that even simple organisms can store molecular memories of past threats and deploy them strategically. The distinction between deliberate genetic intervention and natural evolutionary strategy now appears increasingly thin.

Alongside these genetic revelations, biology is experiencing a renewed focus on form and development. Research on developmental plasticity shows how a single genome can produce radically different outcomes depending on environmental conditions. The African killifish offers a striking example. Its embryos can enter a state of suspended animation, known as diapause, surviving for months or even years in dry mud until rains return. During this period, gene expression shifts dynamically, allowing the organism to switch survival strategies as conditions change. Genes, in this light, function less like fixed instructions and more like scripts awaiting interpretation.

This reimagining of genes as flexible rather than deterministic has been reinforced by epigenetics. Gene expression can be modified not only by mutations but by lived experience. Stress, trauma, nutrition, and toxin exposure can leave molecular marks on DNA that sometimes persist across generations. A famine suffered by one generation may reappear as altered metabolic health in the next.

Even neuroscience is beginning to catch up. The brain, once thought to be largely fixed after childhood, is now known to retain the capacity for adult neurogenesis and profound plasticity. In songbirds, neurons regenerate seasonally as songs are learnt and forgotten. In humans, practices such as meditation can measurably alter the thickness of specific brain regions. Thus, we can conclude that cognition is not a sealed process confined to the skull, but a porous exchange shaped by environment, attention, and emotion.

Then there is synthetic biology. With advances in genome writing, we are no longer limited to reading life; we can now compose it. Cells can be engineered to sense toxins, manufacture drugs, or self-destruct under specific conditions. However, this shift raises unsettling questions. What does it mean to be alive when life itself is programmable? Should life be defined by its origin, its behaviour, or its potential? If we assemble a synthetic cell from basic components, are we creators or merely custodians?

Yet the deepest challenge may not be technical but perceptual. Modern biology asks us to learn to see life not as a finished object but as a continual becoming. Most importantly, it offers an unfinished symphony, where every cell is a stanza, every mutation a metaphor, and every breath a bridge between what has been and what might be. And in this dance of genes and microbes, of growth and decay, of memory and mutation, life persists—not because it is tidy but because it is tenacious; not because it is perfect but because it is possible.

> 'Not only is the universe stranger than we imagine, it is stranger than we can imagine.'
>
> — **Arthur Stanley Eddington**

chapter 6

cosmic smoke and mirrors

Since the earliest days of human thought, the night sky has served as a canvas for imagination, wonder, and inquiry. Our ancestors traced constellations across the darkness, weaving stories of gods, fate, and cosmic order into the stars themselves. Over time, myth gave way to measurement. Telescopes replaced naked eyes, equations supplanted epics, and the universe came to seem—at least briefly—legible. With ever more powerful instruments, we believed we were closing in on the deepest truths of the cosmos, the vast stage on which our fleeting lives unfold.

Yet the opposite has happened. The farther we peer into space and time, the less stable our understanding becomes. Each new layer of clarity reveals a deeper strangeness beneath it, as though reality were performing a carefully rehearsed illusion, always withholding its final move. It is within this growing sense of cosmic uncertainty that two of the universe's most unsettling characters step into view: dark matter and dark energy. They are unseen forces that make up the silent majority of the universe, accounting for about 95 per cent of its total content. They neither shine nor sparkle as they do not interact with light. They cannot be held in our hands and remain beyond full comprehension. Yet they are everywhere, shaping galaxies and driving the cosmos to expand at an ever-increasing rate. In simple terms, they are the universe's shadow play: powerful, elusive, and profoundly mysterious.

Dark matter first emerged as a puzzle when astronomers noticed something strange about the way galaxies behave. According to the laws of physics as we understand them, stars at the outer edges of a galaxy should orbit more slowly than those near its centre, much like a ball on a string that slows when the string is loosened. But observations revealed the opposite. Stars far from galactic centres move just as fast, sometimes even faster. By ordinary gravity alone, such galaxies should fly apart. Instead, they hold together. This implied that something unseen must be exerting additional gravitational pull. Dark matter is the name given to this invisible influence, as if a hidden framework were scaffolding galaxies and clusters from within.

It is important to note that we infer dark matter's presence from its gravitational effects, but what it actually is remains unknown. Scientists have proposed a host of exotic particles—WIMPs, axions, sterile neutrinos, and others—yet none have been conclusively detected. Some researchers suggest that dark matter may interact with itself in unfamiliar ways. Others wonder whether gravity behaves differently at vast cosmic scales than our current theories predict. Each new hypothesis adds depth to the mystery.

While dark matter binds the cosmos together, another invisible force appears to be pulling it apart. In the late 1990s, astronomers studying distant Type Ia supernovae made a startling discovery: The universe is not only expanding but expanding at an accelerating rate. Some unknown force is pushing galaxies apart and stretching space itself. This phenomenon was named dark energy, and it appears to account for roughly 68 per cent of the universe's total energy content. And in order to better understand its implications, we must shift our perspective on the very nature of space.

Dark energy might represent a constant property of space itself, an intrinsic energy woven into the fabric of the cosmos. Or it could be dynamic, changing over time like a cosmic tide. Whatever its nature, dark energy compels us to rethink emptiness. It also highlights the fact that space is not a barren void but a restless arena. Quantum physics tells us that even 'empty' space teems with fleeting particles that flicker into and out of existence. Dark energy's expansive push reinforces a paradox: emptiness is not nothingness. It has structure, tension, and consequence.

Ultimately, the discoveries of dark matter and dark energy unsettle absolute thinking, revealing a universe that is relational, incomplete, and far stranger than it seems. These phenomena are not mere technical puzzles; they challenge our assumptions about existence, showing that most of the cosmos lies beyond direct perception while quietly shaping the world we see.

So when you look up at the night sky, remember that the stars represent only a small fraction of what is out there. The cosmos is larger, stranger, and more dynamic than it appears, and what we call empty darkness is not absence, but a presence waiting to be understood.

Living with Cosmic Uncertainty

In 2025, the Dark Energy Spectroscopic Instrument (DESI) reported results that sent quiet tremors through cosmology. Its data suggested that dark energy, long assumed to be constant and responsible for a steady acceleration of the universe's expansion, may not be constant after all. The findings hinted that this mysterious force could be weakening over time. If that interpretation holds, the consequences are profound. Instead of expanding forever into a cold, dilute future, the universe might one day slow, perhaps even reverse course in a dramatic collapse known as the Big Crunch. Other scenarios are even more dramatic, including the Big Rip, where an increasingly dominant dark energy tears apart galaxies, stars, and eventually matter itself. This possibility does more than revise a parameter in an equation. It unsettles the linear story we had grown comfortable with and replaces it with something more cyclical, where endings may fold back into beginnings.

Unlike Einstein's static vacuum energy, quintessence describes a field that changes over time, its strength shifting as the universe evolves. It offers a more flexible framework for explaining cosmic acceleration, though it remains hypothetical. DESI's findings are being examined carefully, but they have already prompted renewed scrutiny of long-held assumptions.

Alongside DESI's results, the European Space Agency's Euclid mission, launched in 2023, has begun mapping billions of galaxies across cosmic history. Its early data have revealed subtle irregularities in the large-scale distribution of matter, raising questions about the completeness of the standard lambda–cold dark matter (ΛCDM) model. Euclid's maps suggest that the universe's architecture may be more intricate than previously thought, with dark matter and dark energy possibly interacting in ways our current framework does not fully capture. Meanwhile, the Vera C. Rubin Observatory in Chile has begun surveying the sky in unprecedented detail, tracking how billions of galaxies cluster and move. The more precisely we measure the distribution of matter, the better we can infer the behaviour of the unseen force shaping it.

Meanwhile, a 2024 study re-examining Type Ia supernovae, the 'standard candles' used to measure cosmic expansion, introduced another twist. The researchers argued that inhomogeneous cosmological models, which take into account uneven distributions of matter on large scales, might explain the apparent acceleration of the universe without invoking dark energy at all. In this view, what appears to be acceleration could partly arise from how we interpret light travelling through a lumpy cosmos. This means that the expansion may not be an illusion, but our understanding of its cause might be incomplete.

At the same time, an older and once-marginal idea has regained attention: the dark fluid hypothesis. This model proposes that dark matter and dark energy are not separate components but different expressions of a single underlying phenomenon. Rather than two ingredients in the cosmic inventory, they may be two aspects of one unified substance. Such a view challenges the dualistic framework that has shaped modern cosmology and points towards a more integrated understanding of the universe.

More broadly, some theorists are exploring models in which gravity, spacetime, and even quantum fields emerge from deeper informational structures. In these approaches, familiar categories such as mass, force, and energy may not be fundamental entities but effective descriptions of more basic processes. Theoretical physicists are racing to keep pace. Some are exploring whether string theory, often described as a leading candidate for a 'theory of everything,' might hold clues. In certain models, dark energy emerges naturally from the intricate geometry of extra dimensions that string theory requires. Other approaches turn to quantum gravity, where the fundamental structure of space and time itself could generate the vacuum energy we interpret as dark energy. A few researchers have even suggested that the universe's apparent acceleration may reflect gaps in our understanding of time within general relativity rather than a new cosmic ingredient altogether. Thus, the distinctions we treat as absolute could turn out to be provisional.

These conceptual tensions surfaced clearly at a 2023 symposium on the philosophy of dark energy, where physicists and philosophers asked a deceptively simple question: What is dark energy? Is it an intrinsic property of empty space? A new fundamental field? An emergent effect of quantum entanglement? Or is it a placeholder, a term that captures the limits of our current theories?

The astonishing thing is that there were no definitive answers, and perhaps that is fitting. As one participant remarked:

> 'Dark energy may be less a thing than a finger pointing at our limitations.'

What emerges most clearly from these discoveries is that the universe resists being pinned down and does not submit easily to absolutes. Each time we think we have carved reality into clean, Euclidean forms, the edges blur and the corners soften. The cosmos seems to resist reduction, inviting not conquest but encounter. And this is not merely a philosophical stance as the data compel it. The story of dark matter and dark energy, which together account for more than 95 per cent of the cosmos, calls us towards careful inquiry rather than confident finality. We are urged to live with questions and not rush towards premature certainty.

There is a Sanskrit term, anirvacanīya, meaning 'that which cannot be definitively expressed.' In Vedantic thought, it describes realities that are neither fully real nor unreal, neither wholly graspable nor completely absent. The dark universe seems to occupy a similar space. It is detectable through its effects yet elusive in its essence, present in influence but hidden in form.

Recognising this, we begin to see that understanding the cosmos may not always be about uncovering absolute truths. Perhaps the greatest lesson of the dark cosmos is not a new law but a new attitude, not a final answer but a refinement of the questions we dare to ask, reminding us that some of the most powerful truths arrive wrapped in paradox.

Informational Dark Cosmos

In 1948, while television audiences gathered around sets to watch The Ed Sullivan Show, a quieter revolution was unfolding at Bell Labs. Claude Shannon introduced a mathematical theory of information. He showed that information is not just a meaning or a message, but a measurable structure. Put another way, it is order extracted from randomness. With this, information became a quantifiable physical concept.

This perspective casts the cosmic mystery in a different light. As discussed earlier, roughly 95 per cent of the universe is attributed to dark matter and dark energy. Yet 'dark' may be a misleading term. It suggests the absence of light, when the deeper issue may be the absence of accessible information. In other words, the darkness may not be emptiness but encryption.

To better understand this, let's consider a simple analogy. When you delete a photograph from your phone, it does not immediately vanish. The data remains stored until overwritten. The image is invisible to you, but it still exists in coded form. Could something similar be true of the universe? Could dark matter and dark energy represent information present in reality but encoded in ways we do not yet know how to read?

This line of thought gains weight from work done in the 1960s by physicist Rolf Landauer. He demonstrated that erasing a single bit of information requires energy. This established that information goes beyond abstract bookkeeping and can directly influence physical reality. Most importantly, Landauer's principle linked computation, thermodynamics, and energy in a fundamental way.

So, if information has energy costs, then cosmic processes that store, erase, or transform information might have energetic signatures. Dark energy, in this view, may not be a mysterious substance pushing space apart. It could reflect the universe's own informational dynamics, a large-scale accounting of how information is processed and redistributed across spacetime. This is not a finished theory but a reframing. Rather than asking only what dark matter and dark energy are made of, we might also ask what they reveal about structure, entropy, and hidden order.

Within this information framework, we also encounter the holographic principle, which we discussed in earlier chapters. This principle is based on the idea that a hologram encodes a three-dimensional image on a two-dimensional surface. Extending this concept, some physicists have proposed that our three-dimensional universe may likewise be encoded on a distant, lower-dimensional boundary, suggesting that what appears vast and volumetric could, at a deeper level, be fundamentally informational. Research by physicists such as Patrick Hayden and John Preskill shows that such holographic encoding can protect information in a manner strikingly similar to quantum error correction—a discovery that connects cosmology to quantum computing in unexpected ways.

Moreover, we know that at the smallest scales, the universe already defies the notion of smooth continuity. Quantum gravity research suggests that spacetime may have a discrete structure near the Planck length, around 10^-35 meters. This does not mean the universe is literally a video game, but it does raise the possibility that spacetime has an informational grain. In that sense, John Wheeler's phrase 'it from bit' captures an intuition shared by many theorists: physical reality may arise from underlying informational processes. Over the past few decades, many physicists—including Vlatko Vedral and others—have begun to treat information not as a convenient description, but as something fundamental. In this view, the universe can be seen as quantum information in motion, with particles, forces, and even spacetime emerging from patterns of qubits and their entanglement.

Here, thermodynamics adds another layer of complexity. The second law tells us that entropy, often associated with disorder or information spread, increases over time. Some researchers have speculated that cosmic expansion could be related to how information and entropy are distributed across spacetime. Rather than imagining dark energy as a mysterious pressure, one might ask whether it reflects deeper constraints tied to entropy and information flow.

Even the near-flat geometry of the universe has inspired informational interpretations. Cosmological observations show that space is remarkably close to flat on large scales. From a purely physical standpoint, this reflects conditions set in the early universe. From an informational perspective, some have wondered whether flatness corresponds to a state of maximal stability or consistency in how information is encoded across spacetime. These ideas remain speculative, but they illustrate a broader shift in thinking.

To see how abstract principles can manifest in concrete systems, consider a complex machine such as the bacterial flagellar motor. Its individual parts seem purposeless in isolation, yet together they generate motion. In a similar spirit, dark matter might represent unseen components of a deeper structure. We infer their presence from large-scale gravitational effects, even if we do not yet understand their fundamental nature.

Quantum information theory introduces yet another intriguing possibility. In quantum computing, error-correcting codes protect fragile information from noise. Some theoretical physicists have proposed that spacetime geometry itself may reflect a kind of quantum error correction, preserving consistency in the face of quantum fluctuations. These ideas are still being developed, but they suggest that information may not merely inhabit the universe; it may help stabilise it. The most powerful quantum error-correcting codes display a property known as strong contextuality, meaning their informational behaviour cannot be reproduced by classical systems. This supports the idea that the universe's deep structure is not only resilient but irreducibly quantum.

The anthropic principle also takes on a different tone in an informational universe. Instead of invoking special design or an infinity of random universes, one might imagine that certain large-scale properties emerge naturally in systems capable of sustaining complex information processing. In such a view, dark matter, dark energy, and ordinary matter would be different aspects of a single, evolving informational structure.

Bringing these ideas into the realm of observation, researchers are actively testing the dark sector. At places like Fermilab and the University of Chicago, researchers are building increasingly sensitive detectors to search for faint signals from dark matter candidates. Advances in quantum sensing have improved both speed and precision, allowing experiments to explore previously inaccessible regimes. While no definitive detection has yet been made, the technological progress is real.

It is essential, however, to separate scientific possibility from metaphysical conclusion. The suggestion that information underlies physical reality does not automatically imply that consciousness is fundamental to the cosmos or that the universe is literally a computer running a program. Those interpretations go beyond current evidence. What can be said with confidence is that information, entropy, and quantum correlations now occupy a central place in theoretical physics.

Seen this way, the dark cosmos is not merely a storehouse of exotic particles or mysterious energy fields. It may also be a sign that our conceptual tools are still incomplete. Whether dark matter consists of new particles, modified gravity, or deeper informational structures remains an open question. In the end, the dark cosmos is an invitation to refine our theories, sharpen our instruments, and rethink what we mean by matter, energy, and perhaps even reality itself.

The Hidden Sector

Over the course of our exploration, it has become evident that the universe is not only vast but also reticent. It reveals little, conceals much, and when pressed with our most refined instruments, offers only a carefully filtered version of itself. What we know about the cosmos—from the motion of galaxies to the behaviour of subatomic particles—comes to us through the limits of perception, theory, and technology. Yet as our equations grow more elegant and our detectors more sensitive, a persistent absence lingers at the edge of explanation. That absence now has a name: the hidden sector.

This idea is not about black holes or distant nebulae. Nor is it simply a rebranding of dark matter as inert scaffolding holding galaxies together. The hidden sector proposes something more radical: a parallel set of particles and forces that may be as rich and dynamic as the visible world, but that interact with us only faintly, if at all. It is less a shadow than a neighbouring framework, overlapping our own yet largely inaccessible. Recent research suggests that these hidden realms might be understood not simply as collections of undiscovered particles, but as complex informational systems in their own right. From this standpoint, subtle glitches or irregularities in the cosmic code could shape the behaviour of the visible universe.

It is worth noting that this concept did not arise from mysticism but from mathematics. As established previously, the Standard Model of particle physics, despite its extraordinary precision, is incomplete. It does not incorporate gravity and struggles to explain neutrino masses. It also offers no convincing candidate for dark matter, which clearly shapes the large-scale structure of the universe. Rather than endlessly modifying the existing framework, some physicists have proposed an additional realm of matter and force that communicates with our own only through subtle channels.

In this hidden sector, there could be particles analogous to those we know: hidden photons, dark Higgs bosons, even forms of dark atoms and dark chemistry. But these entities would follow their own internal rules, governed by symmetries distinct from those of the Standard Model. The architecture might echo ours in form while differing entirely in substance. So, if the universe is fundamentally a quantum information processor, then hidden sectors might function as informational subsystems, operating under their own internal logic. Understanding them would require more than telescopes and particle accelerators; it would demand a deeper grasp of information itself.

If such a sector exists, there may be narrow bridges between the two realms. Physicists call these bridges portals. One example is the Higgs portal, in which the Higgs boson could interact with a hidden counterpart, producing small but measurable deviations in its decay patterns. Another possibility is the dark photon, a hypothetical particle that mixes weakly with the familiar photon, allowing faint communication between sectors.

In fact, some models go even further. Mirror matter theories propose a complete duplication of the Standard Model, with every particle possessing a mirrored partner governed by mirrored symmetries. This would not be a simple inversion but an entire parallel structure, potentially with its own forms of chemistry and structure formation. It would not interact electromagnetically with us, and so would remain invisible except through gravity. Here, it is essential to note that these are concrete theoretical proposals now under experimental investigation.

Other versions of the hidden sector allow dark particles to interact among themselves. Such self-interactions may help resolve longstanding cosmological puzzles, including why dark matter appears less densely concentrated at the centres of galaxies than simple models predict. If dark matter can redistribute energy internally, it could alter the growth and evolution of cosmic structures in ways that leave gravitational signatures without emitting light. In more speculative scenarios, so-called cannibal dark matter describes hidden particles that interact strongly with one another, forming a subsystem that evolves through distinct thermodynamic and informational regimes.

Additionally, the continued absence of direct detection only deepens the question. Our experiments have not yet captured definitive evidence of hidden particles. This may indicate that the particles are extraordinarily elusive and that our detectors are not tuned to the right signals. Or it may point to a more profound gap in our conceptual framework.

To explore what might lie beyond, some theorists have proposed that this hidden realm possesses its own analogue of the strong nuclear force, sometimes called dark QCD. Just as the visible world binds quarks into protons and neutrons, the hidden sector might assemble dark baryons, dark nuclei, perhaps even a shadow chemistry. Proposals involving hidden glueball dark matter—bound states of gauge fields in a hidden pure gauge theory—suggest that these invisible sectors may possess rich internal dynamics shaped by their own underlying informational architecture. Imagine a dark periodic table that is completely invisible to electromagnetic detection yet evolving alongside us with its own structure and history. Some models even propose a parallel dark nucleosynthesis in the early universe, a hidden genesis whose remnants we cannot see but whose gravitational imprint we still feel. Far from inert, this sector could be dynamic, with its own phase transitions, plasma states, and internal symmetries.

This framework is not mere philosophical speculation. It is grounded in active scientific research. A 2017 study of string theory's hidden sectors, for example, shows that entire families of particles could exist in parallel realms that interact only faintly with the visible universe. A 2019 paper proposes a 'leak-in' mechanism in which dark matter arises from tiny transfers of energy or information from a hidden sector into our own, triggering shifts in that sector's internal state. Other work explores the possibility that hidden sectors undergo phase transitions, much like matter shifts between solid, liquid, and gas. In informational terms, these would resemble different operational modes of a complex computational system, each with distinct physical consequences.

More intriguingly, the visible and hidden realms may share a common origin. In asymmetric dark matter models, the same imbalance that favoured matter over antimatter in the early universe may have produced a corresponding asymmetry in the dark sector. Our existence, and any dark counterpart to it, would then arise from the same primordial irregularity, a slight skew in cosmic bookkeeping that left behind both luminous matter and its unseen partner. Put simply, the hidden sector would not simply coexist but also carry the memory of that ancient asymmetry.

What stands out here is that if the hidden sector mirrors the Standard Model particle for particle, as proposed in mirror matter frameworks, it might support not only physics but also chemistry. And where there is chemistry, the possibility of biology cannot be dismissed outright. Entire life-worlds could, in principle, unfold in this parallel domain, composed of mirror atoms and governed by mirror interactions. Such beings would not be extraterrestrials in the conventional sense. They would be co-inhabitants of the same cosmos, sharing space but not electromagnetic substance. And though our instruments would neither see nor hear them, their presence would be woven into the larger gravitational fabric of reality.

The deeper philosophical provocation of the hidden sector lies in how it unsettles the assumption that reality is exhausted by what we can measure. It suggests that what resists detection may still exert influence, forcing us to reconsider what we mean by matter, interaction, and even existence. If something does not couple to light, is it less real? If it leaves only gravitational traces, does that diminish its ontological status? And if entire domains remain beyond direct access, what does it mean to claim knowledge of the whole?

These are not questions detached from physics. Developments in quantum gravity and string theory increasingly point towards a layered reality in which our familiar four-dimensional spacetime emerges from deeper structures. Some approaches describe the universe as a holographic projection arising from a more fundamental informational substrate, suggesting that the hidden sector is not an optional extension but part of the underlying architecture. Taken together, these ideas portray the universe as a layered and dynamic information system. Visible matter may be only one level within a hierarchy of informational realms. Hidden sectors could function as distinct subsystems governed by quantum codes and error-correcting structures, with their irregularities appearing as subtle anomalies in gravity, particle behaviour, or cosmic expansion.

What makes this moment compelling is not speculation alone but convergence. Theoretical frameworks have matured while experimental tools grow ever more precise. Ultra-sensitive detectors, cryogenic experiments, and high-intensity accelerators now probe the subtle portals through which hidden particles might reveal themselves. Deep underground detectors search for rare dark matter interactions. Cosmological surveys measure the expansion of the universe with increasing precision. Particle experiments probe ever higher energies and finer effects. At the same time, advances in quantum technologies allow researchers to simulate complex informational systems in the laboratory, offering new ways to model the quantum structure of spacetime itself. In the coming decades, the question may shift from whether a hidden sector exists to how its signals eluded us for so long.

This is not a tour of settled knowledge but an invitation to the frontier. The hidden sector may be the silent partner of the visible universe, shaping structure without sharing light. In this emerging view, the cosmos is more than a collection of matter and energy; it begins to resemble an immense quantum information process, with unseen layers influencing the visible world through patterns we are only starting to decode. If so, the universe is not merely larger than we imagined but more layered, more intricate, and more elusive than our current categories allow. The dark cosmos urges us to rethink reality as something woven from quantum information and organised by subtle constraints, a universe whose deepest mysteries may reside not in distant stars but in the logic of the code that underwrites them. Any glimpse into that deeper architecture could reshape not only physics but our very understanding of what is real.

When Constants Start to Wander

For centuries, we have held on to a comforting idea: the universe is ruled by constants. These are the cosmic laws and numbers we take for granted—the speed of light, the gravitational constant, the charge of the electron. They serve as the universe's fixed scorecard, the stable rules of the cosmic game. Light travels at exactly 299,792,458 meters per second. Gravity pulls with a strength encoded in a universal constant. The electron carries a tiny, precisely balanced charge that holds atoms together. These values have acted as science's North Star, guiding everything from GPS satellites to the fusion fires inside the Sun. But what if these constants are not truly constant? It sounds like science fiction, yet it is one of the most provocative questions in modern physics. Some recent research suggests that these supposedly fixed numbers might shift ever so slightly across vast stretches of space and time.

Consider the fine-structure constant. This subtle number governs the strength of electromagnetic forces. In other words, it is the glue that holds atoms together and powers everything from lightning to your phone screen. Around 2010, astronomers studying light from distant quasars, brilliant beacons billions of light-years away, noticed something unexpected. The atomic fingerprints embedded in that light hinted that the fine-structure constant might differ slightly from one region of the universe to another. It is as though nature is whispering the rules differently in different cosmic neighbourhoods.

If confirmed, the implications would be profound. A non-uniform fine-structure constant would mean that the chemistry underlying life might be a local feature rather than a universal rule. Such a possibility raises unsettling questions. Could life elsewhere operate under a different chemistry? Might parts of our physics textbooks need to be rewritten?

And the fine-structure constant may not be the only parameter under scrutiny. The gravitational constant, the quiet force that makes apples fall and planets orbit, has also entered the discussion. In recent years, studies of galaxy clusters and pulsars have raised the possibility that gravity's strength may have differed billions of years ago. The idea that gravity could vary across time or space sounds radical. Yet it offers a potential way to address long-standing cosmological puzzles, including why the universe's expansion is accelerating rather than slowing under gravity's pull.

Then there is the electron's charge, the fundamental unit that makes chemistry possible. So far, it appears stable. However, ultra-precise atomic clocks and quantum sensors are pushing measurements to unprecedented accuracy. Even the tiniest deviation could point towards hidden dimensions or new physics beyond the Standard Model.

The proton-to-electron mass ratio, the number around 1836 that shapes atomic structure, has not escaped attention either. In 2020, astronomers using the Very Large Telescope in Chile reported minute shifts in this ratio by analysing quasar light from billions of years ago. If confirmed, such variations would suggest that even the architecture of atoms may not be identical everywhere in the cosmos. The account does not end here. The weak nuclear force, responsible for radioactive decay and neutrino behaviour, is also being called into question. In 2021, researchers examining ancient minerals reported hints that this force may have behaved differently over geological timescales, though the evidence remains tentative. Additionally, some cosmologists have even explored the bold idea that the speed of light itself may have been different in the early universe. A varying speed of light could help explain enduring mysteries, such as why the universe appears remarkably uniform in all directions.

Another interesting point to note is that questioning the constants of the universe is not a recent phenomenon; it has been a subject of inquiry since the early days of modern science.

Back in 1937, Paul Dirac proposed a bold idea: what if some of the universe's fundamental constants are not truly constant? He suggested that the gravitational constant, G, might slowly change over time. This large numbers hypothesis was his attempt to explain why gravity is so much weaker than the other fundamental forces. Perhaps G is not fixed, he argued, but evolves gradually.

Richard Feynman later examined this possibility using evidence from geology and observations within the solar system. He showed that if G had varied as much as Dirac proposed over the past four billion years, we would already see unmistakable consequences in planetary motions and earth's history. If G changes at all, the variation must be extraordinarily subtle or coordinated with changes in other constants.

This leads to a crucial clarification. Physicists do not examine constants in isolation. Instead, they focus on dimensionless quantities, ratios of physical constants that do not depend on human-made units. The fine-structure constant is a prime example. It combines the electron charge, the speed of light (c), Planck's constant, and vacuum permittivity into a single pure number. If the speed of light changed but the electron charge shifted in precise proportion, leaving the fine-structure constant untouched, no measurable physics would change. That is why researchers test the stability of these dimensionless constants. Only when such ratios vary would the laws of physics themselves truly shift.

In 1957, Robert Dicke explored a related possibility. He developed a framework in which the speed of light might vary over cosmic time, even though it remains locally constant in Einstein's relativity. Dicke suggested that as the universe expands and its large-scale structure evolves, the effective speed of light could slowly decrease. His model accounted for the observed bending of light by gravity and offered an alternative explanation for cosmological redshift, the stretching of light from distant galaxies.

Since then, variable speed of light theories have been refined by researchers such as Jean-Pierre Petit, John Moffat, Andreas Albrecht, and João Magueijo. These models aim to address major cosmological puzzles, including the horizon problem, which asks why distant regions of the universe appear so similar despite having had no apparent time to exchange information. By allowing the speed of light to vary under certain conditions, VSL proposals offer an alternative to relying solely on cosmic inflation.

Taken together, these ideas suggest that the universe's so-called constants may not be carved in stone. Yet any meaningful change must reveal itself in dimensionless ratios, the quantities that truly govern physical law. This discipline keeps scientific tests precise while leaving open the possibility that the cosmos is more dynamic than once imagined. Some of the most ambitious modern theories predict exactly this kind of variability. For instance, in string theory, extra spatial dimensions are curled up at unimaginably small scales. The way these dimensions are shaped could subtly influence the values of observable constants.

The multiverse hypothesis pushes the idea further. It proposes a vast ensemble of universes, each with its own set of physical constants. In that view, what we call constants may be local rules that hold within our cosmic region but differ elsewhere. Our universe would then be one outcome in a grand statistical spread, its life-permitting values less a miracle than a selection effect.

It is also important to recognise that these inquiries are not merely philosophical. Even minute changes in fundamental constants would alter how stars burn, how atoms bond, and how galaxies evolve. They would influence the long-term fate of the universe: endless expansion, catastrophic tearing, or eventual collapse, all depend on how these numbers behave over cosmic time.

New instruments are now probing these possibilities with unprecedented precision. The Dark Energy Spectroscopic Instrument is mapping the expansion of the universe in extraordinary detail. The European Space Agency's Euclid mission and the Vera C. Rubin Observatory are sharpening our picture of cosmic structure and evolution. Their measurements may confirm that certain constants drift over time, or they may constrain such variations so tightly that entirely new ideas are required.

Meanwhile, laboratory experiments on earth are reaching astonishing sensitivities. Ultra-precise atomic clocks, quantum sensors, and interferometers test whether quantities such as the fine-structure constant vary ever so slightly with time. We are entering an era in which the smallest measurable intervals and the largest observable scales are examined together, searching for subtle departures from permanence.

Roger Penrose once remarked that the laws of physics display an extraordinary elegance. The question now is whether that elegance lies in rigid immutability or in a deeper, evolving order we are only beginning to glimpse.

> '[T]here's a version of this, a version of this idea which John Wheeler has promoted, which is that in each of these cycles, since nobody really knows what goes on at the crunch,

> bang stage . . . you can . . . invent any physics you like, and one idea . . . is to suggest that the . . . fundamental constants of nature might get changed every time you go through one of these cycles . . . [T]his might help to explain . . . puzzles that . . . the constants have to be just such and such in order that life should exist . . . [etc.] I always have trouble with many of these arguments. It is not at all clear whether you need them or not. They might be true but we don't know. It may be that these numbers are fixed and they might change through each cycle . . . [etc.] but our current physics . . . doesn't allow this kind of thing. These are singular states according to classical theory. Maybe if we had quantum gravity . . . one could imagine such a scheme . . .'

What does all this tell us beyond the equations and the data? It suggests that the universe is stranger and more wondrous than we once imagined. The constants we long treated as fixed may not be immutable laws but evolving parameters capable of change, much like living systems that adapt over time.

In this sense, science appears to be circling back to an older intuition that change, not permanence, is fundamental. The cosmos begins to look less like a rigid machine and more like a dynamic unfolding, inviting us to rethink not only physics but our place within it. We are not merely passive observers but participants in a story still in motion, where even the deepest rules may not be final. It is both a humbling and exhilarating possibility. Wandering constants remind us that the universe remains alive with mystery, urging us to explore, to question, and to remain open to surprise. As those constants shift, however subtly, so too does our understanding, moving and growing along with the cosmos itself.

Cosmic Quicksand

What if the universe beneath our feet is not solid ground but cosmic quicksand, stable enough to stand on today yet slowly shifting and capable of upending everything we know? Welcome to the strange realm of vacuum decay, where the foundations of reality may not be as permanent as we once assumed.

At its heart, vacuum decay is the unsettling possibility that the vacuum we inhabit—the very fabric of empty space that determines particle masses and the strengths of forces—might actually be a false vacuum. Picture a cosmic bubble that appears stable but sits in a shallow valley, while a deeper and more stable valley lies close by. If our universe were to tunnel into that true vacuum, it would trigger a dramatic phase transition, rewriting the laws of physics in an instant.

This may sound like science fiction, but it is grounded in serious physics. In recent years, experiments at the Large Hadron Collider, along with increasingly precise astrophysical observations, have sharpened our understanding of the Higgs boson and the top quark—two crucial players in shaping the vacuum's stability. Detailed studies by Degrassi, Buttazzo, and others suggest that we may be balanced on a razor's edge between stability and metastability. Our cosmic bubble is unlikely to burst anytime soon—on timescales far beyond trillions of years—but it is not guaranteed to endure forever.

The story grows even more intriguing when we consider the hidden sector, populated by dark matter and possibly other exotic fields. These unseen ingredients could subtly reshape the Higgs landscape, making vacuum decay either more or less likely. Invisible stagehands may be quietly influencing the fate of the cosmic performance.

This perspective resonates with contemporary interpretations of quantum physics, such as Carlo Rovelli's relational quantum mechanics, where reality arises from interactions rather than fixed, independent objects. In this view, the universe becomes a dynamic, reflexive system in which information, fields, and quantum fluctuations intertwine in a continual process of becoming.

Physicists like Max Tegmark have gone further, suggesting that the universe's large-scale structure and even its fundamental constants may emerge from deeper mathematical and informational patterns. On this reading, the boundary between physics and philosophy begins to blur. Vacuum decay also carries an existential undertone: existence itself may be provisional, a fleeting configuration in a cosmos defined by change and possibility. This stands in stark contrast to the old Newtonian vision of a perfectly ordered clockwork universe running smoothly and eternally.

As Sidney Coleman and Frank de Luccia famously observed:

> 'The possibility that we are living in a false vacuum has never been a cheering one to contemplate. Vacuum decay is the ultimate ecological catastrophe; in the new vacuum, there are new constants of nature; after vacuum decay, not only is life as we know it impossible, so is chemistry as we know it.'

Some physicists have also suggested that the rapid expansion of the early universe—cosmic inflation—may have been triggered when the Higgs field escaped from a metastable false vacuum. In this picture, the Higgs field, the invisible agent that gives particles their mass, was trapped in a state that appeared stable but was not the lowest possible energy configuration. A small quantum fluctuation could have pushed it into a more stable state, unleashing an enormous burst of expansion.

At the tiniest scales, near the Planck scale, two crucial properties of the Higgs field—its self-coupling and the rate at which that coupling evolves, known as the βλ function—appear to hover close to zero. This delicate balance suggests that the Higgs potential may sit near a critical boundary between stability and instability. The image is almost that of a tightrope walker paused mid-step, balanced between firm footing and a sudden fall. And to determine which side of that boundary we occupy, physicists are looking to future electron–positron colliders. These next-generation accelerators are designed to measure the mass of the top quark with extraordinary precision. That single value plays a decisive role in calculating the stability of the Higgs field. A slight shift could mean the difference between a universe resting safely in a stable vacuum and one perched in a long-lived but ultimately metastable state.

Alan Guth, who first proposed cosmic inflation, envisioned the early universe emerging through quantum bubble nucleation, like a superheated liquid suddenly boiling over. He famously said:

> 'During inflation, as the universe gets bigger and bigger and more and more matter is created, the total energy of matter goes upward by an enormous amount. Meanwhile, however, the energy of gravity becomes more and more. The negative gravitational energy cancels the energy in matter, so the total energy of the system remains whatever it was when inflation started—presumably something very small . . . This capability for producing matter in the universe is one crucial difference between the inflationary model and the previous model.'

Guth's original picture of quantum bubble nucleation involved a dramatic tunnelling event through the quantum fabric of space. But such a violent transition would have disrupted the remarkable smoothness of the universe we observe today. To address this, Andrei Linde, Andreas Albrecht, and Paul Steinhardt proposed a more measured alternative called slow-roll inflation. Instead of a sudden leap, the universe evolves gradually, like a skier gliding down a gentle slope rather than plunging off a cliff. This refinement preserves the universe's homogeneity and isotropy, meaning it appears roughly the same in every direction.

There is another twist that often goes unnoticed. In 2014, researchers at the Wuhan Institute of Physics and Mathematics showed mathematically that a universe could spontaneously arise from nothing. Here, nothing does not mean empty space. It means no space, no time, and no matter—only quantum fluctuations within a metastable false vacuum giving rise to an expanding bubble of true vacuum. It is a striking example of how modern physics stretches the meaning of existence itself.

This idea echoes themes from ancient philosophy. In Indian cosmology, *shunya*, or the void, is not mere emptiness but a fertile ground from which reality emerges. In a similar way, the quantum vacuum is not inert absence but a seething potential from which universes can arise. Inflation, then, is more than a technical model. It is a story of transition: from false vacuum to true, from instability to structure, from quantum uncertainty to cosmic order.

Additionally, the mathematics behind vacuum tunnelling relies on objects called instantons. The probability of decay depends on the instanton's action, a quantity that enters an exponential formula determining how likely a bubble of true vacuum is to form. A smaller action means the false vacuum is more susceptible to transition. When gravity is included, the picture becomes richer. Coleman–De Luccia instantons describe how such bubbles nucleate and expand in a curved, inflating spacetime, significantly altering the decay dynamics compared to a flat, gravity-free setting.

Recent computational advances have sharpened these predictions. Work by Espinosa and others indicates that the conditions of the early universe, including inflation itself, strongly influence how vacuum decay proceeds. And although much remains uncertain, the theoretical framework is becoming increasingly precise.

In summary, vacuum decay is not a fringe curiosity but a window into the deeper structure of reality, where hidden sectors and additional fields may help shape the fate of the cosmos. At its core lies a disquieting insight that the laws of nature may not be fixed and eternal but emergent and contingent upon the vacuum state the universe happens to inhabit. And to explore vacuum decay is therefore to confront a universe dynamic at its foundations, where permanence may be an illusion and stability only provisional.

epilogue

dancing with the unknown

As we close these pages, we find ourselves not at an ending, but at the edge of a beginning . . . the beginning of a new relationship with reality itself.

This journey began with a simple, unsettling question: What if our search for absolute truth was not just difficult but fundamentally misguided? What if the universe was not a puzzle to be solved but a conversation to be joined? Across physics, philosophy, biology, and economics, we have encountered the same recurring theme: reality thrives on uncertainty, resists final answers, and delights in holding contradictions in tension. From the quantum haze where particles exist as clouds of probability to the cosmic scales where dark energy accelerates an expanding mystery, the universe refuses to be pinned down.

The great insight of our time is not a new equation or a final theory. It is the recognition that not knowing is not a failure, but a form of wisdom. It is the humility to accept that our maps are not the territory, our models are not the cosmos, and our certainty is often just a story we tell ourselves to feel at home in a vast and strange existence.

This does not mean we abandon reason, science, or the search for understanding. It means we pursue them with open hands, not clenched fists. We learn to think in probabilities, not certainties. To value questions as much as answers. To see knowledge not as a fortress to defend, but as a garden to tend—a living, growing thing that requires curiosity, care, and the courage to be wrong.

It is also vital to realise that the universe is not pranking us out of malice. Instead, it is inviting us to play. To step out of the rigid architectures of absolutism and into the flowing, participatory dance of existence. And when we do this, something remarkable happens. We stop trying to control the mystery and start belonging to it.

So, take this motto with you: the freedom to be uncertain. The courage to hold two truths at once. The grace to live with questions that have no easy answers. The world does not need more gurus selling certainty. It needs more explorers comfortable with wonder.

After all, the cosmos is vast, strange, and beautifully paradoxical. And so, thankfully, are we.

acknowledgements

The explorations comprising *The Universe Is Pranking Us* were made possible through the steadfast support and encouragement of many individuals. I extend my sincere gratitude to my mentors and teachers, whose guidance has been invaluable.

Professor Brian Josephson, Nobel laureate in Physics (1973) and one of the most distinguished physicists of the twentieth century, has been instrumental in shaping my intellectual journey. Our conversations ranged from non-linear dynamics and condensed-matter physics to deeper questions concerning mind–matter relations and the foundations of physical reality.

In a scientific landscape increasingly framed in terms of information, correlations, and emergent structure—where principles and meta-dynamics may be more fundamental than mass and energy alone—we explored the role of coordination and relational order in nature. More recently, these discussions have led us even further, towards the recognition that even such foundational premises may themselves be contingent rather than absolute, a perspective that resonates deeply with the themes of this book.

I also wish to acknowledge Prof. Roger Penrose, Prof. PeterHiggs, Prof. Crispin Barnes, Prof. Dipankar Home, Prof. Sisir Roy, Prof. Prasanta Panigrahi, Prof. Subhash Kak, and Prof. P. C. Deshmukh for the privilege of discussing the foundations of contemporary scientific frameworks, particularly quantum physics. These conversations, whether direct or indirect, have shaped my understanding of how modern science grapples with questions of reality and rationale, context and constraints.

I would like to honour and acknowledge the enduring intellectual legacy of Kurt Gödel, Ilya Prigogine, Richard Feynman, Albert Einstein, David Bohm, Alfred Tarski, Gregory Chaitin, and Edward Witten, whose ideas continue to illuminate the deep structure of mathematics, logic, and physical law. Their work stands as a reminder that the most profound advances in science often arise from questioning what once seemed unquestionable.

This book is indebted not only to their seminal contributions but also to the spirit of intellectual courage they embodied—the willingness to confront uncertainty, to embrace incompleteness, and to explore the boundaries of reason itself. Any insights presented here are informed by that tradition, while any shortcomings remain entirely my own.

My heartfelt thanks go to my wonderful family: Ma, Dr Karabi M. G. Majumdar; Baba, Prof. Rupendra Guha Majumdar; my wife, Devyani Guha Majumdar; and my brother, Tirthankar Guha Majumdar. I also acknowledge the joy and promise that my firstborn, Anandalekha, has brought to me during the writing of this book. My family's unwavering support and patience throughout its creation have been invaluable. Their belief in my vision provided the strength and motivation needed to complete this work. I would also like to honour my forbears, particularly Thakurma, Aita, Dadu, and Koka Deuta.

I offer my heartfelt thanks to my publishers, Hay House India, for their belief in this project and their unwavering support in bringing this book to life. Their professionalism and commitment have ensured that this work reaches a wider audience.

Lastly, I extend sincere appreciation to all readers and seekers who engage with this book. I hope it sparks curiosity and fosters a deeper understanding of the universality and profound significance of non-absolutism in our universe.